WILD
WEATHER
THE TRUTH BEHIND GLOBAL WARMING

WILD WEATHER
THE TRUTH BEHIND GLOBAL WARMING

DR. REESE HALTER

ALTITUDE PUBLISHING

PUBLISHED BY ALTITUDE PUBLISHING LTD.
1500 Railway Avenue, Canmore, Alberta T1W 1P6
www.amazingstoriesbooks.com
1-800-957-6888

Copyright 2007 © Dr. Reese Halter
All rights reserved
First published 2007

Publisher   Stephen Hutchings
Associate Publisher   Kara Turner
Editor   Michelle Lomberg
Proofreader   Frances Purslow
Layout and Design   Bryan Pezzi

Extreme care has been taken to ensure that the information contained in this book is accurate and up to date at the time of printing. However, neither the author nor the publisher is responsible for errors, omissions, loss of income or anything else that may result from the information contained in this book.

All web site URLs mentioned in this book were correct at the time of printing. The publisher is not responsible for the content of external web sites or changes that may have occurred since publication.

In order to make this book as universal as possible, all currency is shown in U.S. dollars unless otherwise stated.

We acknowledge the financial support of the Government of Canada through the Book Publishing Industry Development Program (BPIDP) for our publishing activities.

ALTITUDE GREENTREE PROGRAM
Altitude Publishing will plant twice as many trees as were used in the manufacturing of this product.

For general information on Altitude Publishing, including all books published by Altitude Publishing, please visit our Web sites at www.altitudepublishing.com and www.amazingstoriesbooks.com or call our order line at 1-800-957-6888. For reseller information, including discounts and premium sales, please call our sales department at 403-678-9592. For press review copies, author interviews, or other publicity information, please contact our marketing department at 403-283-7934, or via fax at 403-283-7917.

Cataloguing in Publication Data

Halter, Reese
        Wild weather : the truth behind global warming / Reese Halter.
Includes bibliographical references.
ISBN 978-1-55439-532-3

        1. Global warming. 2. Greenhouse effect, Atmospheric. 3. Climatic changes. I. Title.
QC981.8.G56H35 2007   363.738'74   C2007-900677-9

In Canada, Amazing Stories® is a registered trademark of Altitude Publishing Canada Ltd. An application for the same trademark is pending in the U.S.

Printed and bound in Canada by Friesens
2 4 6 8 9 7 5 3 1

For my wonderful friend and loving wife, LuAn

CONTENTS

1. The Deadly Surge 9
2. Wild Weather. 19
3. Hurricanes. 25
4. Tornadoes. 43
5. Blizzards 57
6. Ice Storms. 73
7. Drought 88
8. Fire 100
9. Global Warming 113
10. Hope for the Future 145

    Wild Weather Timeline 151
    Amazing Facts and Figures 154
    What Others Say 155

CHAPTER 1
THE DEADLY SURGE

As the eye wall of hurricane Katrina approached New Orleans in the early hours of August 29, 2005, the worst predictions of Dr. Walter Maestri, director of Jefferson Parish's office of Emergency Management, were about to come true. The astounding 34-foot (10.4 meter) storm surge easily breached the levees, swamping most of the city of New Orleans and trapping 100,000 people.

Each year, the Hurricane Center at Louisiana State University in Baton Rouge is very busy between June and November — hurricane season — as staff keep watchful eyes on weather maps, satellite data, and computer models.

In 1979, the Federal Emergency Management Agency (FEMA) was created by President Jimmy Carter. FEMA's role is to juggle the vast resources of the federal government, including coordinating and enlisting the services of other agencies, to help Americans when disaster strikes.

In 1981, the director of the National Hurricane Center, Dr. Neil Frank, told *U.S. News & World Report*, "Without question, we've felt through the years that New Orleans is one of the most vulnerable places in the United States. Not only does the area have bad hurricanes fairly frequently, but it has a tremendous concentration of people. Many of those people live in low-lying areas that could be severely flooded. They really don't have a workable evacuation plan for that city."

In the early 1980s, Madhu Beriwal, an emergency planning expert, began working on Louisiana's first hurricane evacuation study.

When asked what was done with the surge models, evacuation times, and public opinion survey information, Beriwal replied, "I found it quite horrifying that there really was no plan for this information. It was not going anywhere, there was a disconnect between the technical scientific information we were producing and operations. There was just no place to put it. No one was looking at how long it took to evacuate and how much water would come from various storms."

During the 1990s, other Louisiana emergency planners began to worry about the lack of hurricane preparedness at local, state, and federal levels. Louisiana State University coastal scientist Dr. Joseph Suhayda acknowledged, "It became clear we had Category 5 storms, Category 3 levees and tropical storm support at the state and local level."

In early 2003, Michael D. Brown, with no previous disaster management experience, was appointed director of FEMA, which after 9/11 became one of the 22 agencies within the newly created Department of Homeland Security.

After hurricane Ivan in 2004, FEMA agreed to co-sponsor a five-day hurricane symposium at Louisiana State University. Beriwal's company would run it and the Hurricane Center's super computer — dubbed SuperMike — would simulate a Category 5 hurricane and its storm surge. The fictitious hurricane was named Pam.

Some months later, more than 300 federal (including FEMA's director, Michael D. Brown), state, local, and volunteer organizations met in Baton Rouge at the university's new Emergency Operations Center and State Police Headquarters to simulate Pam, the Category 5 hurricane.

Pam breached the levees.

"Metropolitan New Orleans was under water. There were 50,000 fatalities and 100,000 casualties. So what are you going to do?" asked Dr. Maestri.

A search and rescue exercise posed the following questions: How many people would be needed? Where would rescuers find boats, and how would they get trapped people off roofs, out of homes, and out of the water? How were the dead bodies to be collected, transported, and stored? How were people with medical problems to be evacuated and cared for? How were the sewage and drainage systems to operate? How were pumps to be rebuilt? How were tons of hazardous materials to be handled? And where was temporary housing to be located?

"FEMA, the state, and local officials are all meeting together and we're all agreeing on what will happen. This is what I'm going to do and what you're going to do — what you will do for me and me for you. Everyone is signing off and saying this is what we can do, and that's what in essence, the

New Orleans, LA, August 29, 2005: An aerial view of a neighborhood with homes flooded to their eaves. In the background is one of the breeched levees with water entering the area.

final report of the Pam exercise spells out. Who's responsible for what and when, and what time span," said Dr. Maestri.

It became clear to many of the local scientists that evacuation of more than a million people was of paramount importance in order to avoid hundreds of thousands of deaths.

Dr. Shirley Laska of the University of New Orleans estimated that at least 134,000 people in New Orleans did not own transportation and 29 percent of the city's population lived in squalor — well below the poverty line. Her main concern was that those impoverished people would resist evacuation.

The Department of Homeland Security, to which FEMA belongs, canceled the funding for Pam soon after the meet-

THE DEADLY SURGE 13

New Orleans, LA, October 17, 2005: An engineer named Tim Fontenot surveys a levee to determine what caused the breach in the Lower 9th Ward following Hurricane Katrina.

ing, and no further opportunities for any additional dry runs were possible.

On Friday morning, August 26, 2005, a thunderstorm that was born three days earlier off the West African coast moved into the Caribbean, became hurricane Katrina, and now was setting its sights on landfall west into Louisiana by Monday.

Max Mayfield, the director of the National Hurricane Center in Miami, watched the satellite information with awe and trepidation as Katrina gained momentum heading into the Gulf of Mexico.

He called Maestri and said, "This is what we've been talking about all these years. My staff think this will track right at you and it has definite potential to become a Category 4 or 5. No, I'm not kidding."

Maestri then called Mayor Nagin, Governor Blanco, and Colonel Jeff Smith, the chief of the Louisiana Office of Homeland Security and Emergency Preparedness. By 4 p.m., Governor Blanco had declared an official state of emergency.

At 7 a.m. on Saturday, Katrina's epicenter was in the middle of the Gulf, with winds in excess of 115 miles per hour (185 km/hr) — it was officially a Category 3 hurricane.

The eye wall appeared for a short while to disintegrate or weaken. Katrina was simply shifting her energy as she began to crank herself up. Within two hours, she had doubled in size and become a Category 4.

Hundreds of thousands of people began leaving New Orleans and the surrounding parishes. State police measured

NEW ORLEANS CROSS SECTION

The lowest areas of the city were the hardest hit, with 80 percent of New Orleans under water.

- **29.5 feet (9 m)**
- **19.7 feet (6 m)**
- **9.8 feet (3 m)** Mississippi River (usual level) — Flood Wall — French Quarter
- **Sea level**
- **-9.8 feet (-3 m)**

Source: The Washington Post

the flow of traffic out of the city at 900 vehicles per lane per hour in the middle of the afternoon. By 7 p.m. it had increased to 1300 vehicles per lane per hour.

By late Sunday afternoon, at least 10,000 people sought shelter in the Superdome stadium. And by nightfall, some 800,000 people had evacuated New Orleans.

There were, however, at least 100,000 who remained in their homes, willing but in reality not able to hunker down against nature's fiercest storm — a Category 5 hurricane — which is equivalent to five times the daily energy output of all the world's power plants.

When Katrina landed in the wee hours of Monday morning, the deadly 34-foot (10.4-m) storm surge easily breached the levees around New Orleans.

At 4 a.m., 50 miles (80 km) southeast of New Orleans in

New Orleans, LA, September 9, 2005: A Blackhawk helicopter drops sandbags into an area where the levee was breached by the Hurricane Katrina storm surge.

the town of Narin, Bobbie Moreau awoke to high-pitched whining — the pressure difference between the inside and outside of her house — interspersed with the most awful creaking noise that she'd ever heard — a wall of Gulf water pressing her house.

She got her daughter, Tasha, her granddaughter Cassidy, and three dogs up to the second floor in what seemed like a split second.

Katrina had landed.

The mother and daughter flopped to their knees. "Please, God, please save us."

Despite raging torrents of water rising feet every second, somehow they managed to squeeze themselves, the four-month-old baby, and three dogs out of the bedroom window onto their slippery tin roof.

Tasha's adrenaline coursed through her body as she went into the water, swam around the house, and found an aluminum flat-bottomed marsh boat. She managed to get the outboard motor going, wheeled the boat around to the roofline and rescued her Mom, her baby, and the dogs.

After a frantic 11 hours of floating without food and fresh water, they were among the very lucky hundred or so to be rescued by a Coast Guard helicopter within the first 12 hours of the arrival of the deadly surge.

The flood trapped at least 100,000 people.

At 5:02 a.m., the power in the Superdome went out. The emergency generators sputtered and then came on but without the air conditioning. Ten thousand people huddled in the Superdome where supplies of food, water, and medicine were desperately low. A public-health disaster was perilously imminent.

America's worst natural disaster was just beginning to unfold.

CHAPTER 2
WILD WEATHER

People are fascinated by weather, and rightfully so. It affects every aspect of our daily lives from what we eat to where we travel. It controls where we live and the fresh water supply upon which we depend for survival. Weather is always on our minds. An entire television station is dedicated to it. Weather has recently demonstrated, and repeatedly demonstrates, its force and violence, ravaging communities and countries. More than just temperature and precipitation, weather comprises a range of events that includes hurricanes, tornadoes, blizzards, ice storms, drought, and fire.

Extreme or wild weather events are those that surpass known records. Those records include physical measurements like wind speed, or temperature, and the amount of damage sustained and its economic cost.

Since the inception of official records, there have been some exceptional extreme weather events, particularly in North America. Some examples are the snowstorm of 1888 that killed more than 100 children; the terrifying tornado in Waco, Texas, in 1953 that killed 114 people; the disastrous ice storm of 1998 that affected tens of millions of people and cost in excess of $7 billion; the epic decade-long drought that is still gripping the American southwest plus the ensuing mega-firestorms of 2003 that plagued the West; and the monster hurricane Katrina that obliterated New Orleans in September 2005.

Scientists have many ways of estimating Earth's temperature and climate from past millennia, including the analysis of tree rings, cores of polar ice, ocean sediments, and lake beds. A glimpse back in time tells scientists that about 2 million years ago, Earth entered the last Ice Age, known more correctly as the Pleistocene epoch.

During the Pleistocene, there were tremendous shifts in climate, and scientists have documented 17 glacial-interglacial cycles. Of these events, seven complete glacial-interglacial cycles have occurred over the past 620,000 years, lasting between 88,000 and 118,000 years.

Those records, however, cannot reveal specific wild weather events, so we must rely upon human records.

North Americans have been keeping official weather records since about 1820. Dutch records extend back to 1706. And the English began keeping weather records in 1659. These records provide a yardstick in helping to determine climatic trends and in measuring the relative severity of weather events. Early records clearly reveal that Earth's temperatures, particularly in the last decade or so, have dramatically risen. Although almost 350 years of weather records easily out-distance a human life span, they are analogous to a geologic

WILD WEATHER 21

One of the thousands of bridges destroyed by Hurricane Katrina.

split-second in the hourglass of the 4.5 billion years of history on our planet, which has undergone ice ages and thaws, whose continents have morphed from tropical to arid, and which is now seeing an unparalleled rise in temperature.

Global Warming

Over the past 50 years, the human population has increased at an unprecedented rate and so, too, has the burning of fossil fuels such as coal, natural gas, gasoline, diesel, and jet fuel. The release of certain so-called greenhouse gases like carbon dioxide (CO_2), methane, and 28 other trace gases, including water vapor, into the atmosphere, has caused Earth's average annual temperature to dramatically rise. This has occurred most notably in the late 1990s and the first part of the 21st century. These rising surface and ocean temperatures are called global warming or climate change. As a result of global warming, some extreme weather events have occurred and more and stronger events are predicted.

The recent accelerated die-off of ancient coral reefs in the Caribbean is an excellent example of global warming and its disastrous ecological effects. As ocean and sea temperatures continue to rise, warmer water, which becomes more acidic, causes the algae that provide food for the coral to die. The coral then succumbs to disease, dies and turns white. This process is called bleaching.

In the past, only some species of coral have suffered bleaching, and die-off has only been recorded at certain water depths. In 2005, in a matter of a few months, bleaching around the island of St. Croix took place at all water depths in the Caribbean Sea, eradicating 96 percent of lettuce coral,

93 percent of star coral, and almost 62 percent of brain coral. Coral reefs are rich and diverse ecosystems depending upon algae to provide essential food for the base of the food chain. In turn, they support all other aquatic flora and fauna. Now, 1,000-year-old coral reefs are dying in their entirety. When this habitat is destroyed, other aquatic plants and animals immediately disappear. People, too, are affected, as the multi-billion dollar tourism and commercial fishing economies are ripe for a fall. "We haven't seen an event of this magnitude in the Caribbean before," said Dr. Mark Eakin, coordinator of the National Oceanic and Atmospheric Administration's Coral Reef Watch.

Further evidence of global warming was the 2005 hurricane season. Hurricanes draw their energy from warm south Atlantic Ocean temperatures. As salt-water temperatures rise, so do the severity, intensity, and frequency of hurricanes. In August 2005, this wrath culminated with Katrina running amuck in Louisiana, Mississippi, and Alabama, inflicting horrendous damages in excess of $75 billion.

"The hurricanes we are seeing are indeed a direct result of climate change and it's no longer something we'll see in the future, it's happening now," said Dr. Greg Holland, a division director at the National Center for Atmospheric Research in Boulder, Colorado.

Global warming has created prolonged droughts on certain continents, including Australia and western North America. Australia is currently experiencing its worst drought in 100 years. In 2002, Arizona experienced its worst drought in 1,000 years. As a by-product of drought and mismanaged wild forests, wildfires are now a serious threat to many communities throughout the West.

Throughout North America and elsewhere, wild forests have evolved to contend with lightning-induced fires. When these natural frequencies of the fire cycle are deliberately altered, serious consequences occur. The prolonged drought in the West has weakened many trees. Fire suppression over the past 80 years has created overstocked, unhealthy forests. As a result, billions of native bark beetles from New Mexico, Arizona, Utah, Nevada, Oregon, California, Colorado, Wyoming Montana, Idaho, Washington, Alberta, British Columbia, Yukon, and Alaska are breeding, feeding and destroying enough drought-weakened trees to feed the entire U.S. housing market for five years. Opportunistic bark beetles have become Mother Nature's emissaries of change, and global warming is fueling the largest ever outbreaks of these insects recorded in modern history.

This book will focus on wild weather, offering contemporary insight into causes of storms and how violent and changing weather affects people, animals, forests, and oceans. It will also examine global warming, explaining why more wild weather is predicted. Everyone cares about weather. It is predicted to change abruptly in the years ahead, making it a topic that needs to be understood.

CHAPTER 3
HURRICANES

Hurricanes are nature's fiercest storms. The energy released in just one hurricane is as much as in 500,000 atom bombs. About 18 hurricanes occur each year. When Bikini Island was demolished by a thermonuclear bomb test in 1954, the explosion lifted about 5.5 million tons of water into the air. That same year, Hurricane Hazel over Puerto Rico drenched the island in 1.37 billion tons of water. In 1969, Hurricane Camille dumped so much rain with such ferocity that it filled the overhead nostrils of birds and drowned them in the trees. In 1970, a typhoon, which is a hurricane that occurs in the Pacific, killed 1 million people in Bangladesh. In 2005, Hurricane Katrina wreaked havoc in Louisiana, Mississippi, and Alabama and left 1.3 million people homeless.

A hurricane is a huge heat pump that gathers the sun's

Tens of thousands of acres were flooded when Katrina came ashore. It took weeks for the lands to drain off.

heat from a large area over the ocean and pumps it into a concentrated region. It warms the air, making it rise, sucking in air from the outskirts to fill the void and forcing the entire air mass to rotate, counterclockwise, faster and faster. Finally, it collapses in on itself. Hurricanes form only over tropical oceans where water temperatures are at least 77 degrees Fahrenheit (25° Celsius). They only occur during the warmest months, from June to about November. They pick up not only heat, but also moisture over the ocean.

The normal path of a hurricane is to the west across the Atlantic and then around to the north as it approaches the continental United States. For a hurricane to be a threat to Miami, it has to start out far to the south below the islands of the West Indies at about 7° latitude. Hurricanes usually do not extend beyond a 30° latitude.

The eye of the hurricane moves very slowly, at about nine miles per hour (15 km/hr), yet its outer wall can have storm-force winds in excess of 186 miles per hour (300 km/hr). That's because in the wall of the eye, ascending winds are lifting air at more than 551,156 tons per second. The average house is built to withstand about 87-mile-per-hour (140 km/hr) winds. In the six hours before and six hours after a hurricane hits land, it can drop over 14 inches (356 mm) of rain. In addition, the storm surge of waves that come ashore can be as high as a three-story building. In Galveston, Texas, in 1900, one such storm surge hit the coast and killed at least 8,000 people.

What conditions create such violence? First, swirling atmospheric conditions that occur off the coast of West Africa create a moist, low-pressure or easterly wave. Then, surface winds from the equator are displaced northward (in the northern hemisphere) and converge with the easterly wave. Third, when a very high-altitude anticyclone (spinning clockwise) sits directly above the center of a low-pressure tropical storm, then all of the necessary ingredients are in place to create nature's greatest storm. The upper-level high pressure area in the center pushes the air away, and the low pressure area at sea level sucks in air and sends it skyward into the center of the anticyclone, which then continues to build up pressure and spin the air away. Warm tropical seawater is evaporated, providing energy, and the subsequent condensation releases heat into the center of the storm as it feeds upon itself.

The problem facing atmospheric scientists today is that they know how hurricanes move but they cannot predict with any certainty, even with the most potent supercomputers, the exact path.

In 1969, the Saffir-Simpson hurricane damage scale, from category one to five, was invented based upon the pressure of the system, its wind speed, and storm surge. Hurricane Charley (2004) was a Category 4 storm with wind speeds in excess of 131 miles per hour (210 km/hr) and 14.8-foot (4.5 m) swells. It decimated Florida's $9 billion citrus industry and destroyed properties worth approximately $7.5 billion.

Fleets of geosynchronous satellites orbit Earth and provide ongoing surveillance over the Atlantic and Gulf of Mexico. They enable scientists to watch the birth and growth of hurricanes from beginning to end. As a hurricane approaches the continental United States, air force airplanes fly into the eye of the storm and collect important information.

THE SAFFIR-SIMPSON HURRICANE SCALE

The Saffir-Simpson Hurricane Scale rates the intensity of a hurricane on a scale of one to five. The rating is used to estimate how much damage a hurricane will cause when it reaches land.

Category	Wind speed	Examples of estimated damage
1	74–95 mph (119–153 kmph)	Some damage to mobile homes, trees, and shrubs; some coastal road flooding
2	96–110 mph (154–177 kmph)	Some damage to roofs, doors, and windows of buildings; some trees blown down; flooding of coastal and low-lying roads
3	111–130 mph (178–209 kmph)	Damage to houses and small buildings; large trees blown down; mobile homes destroyed; small structures near the coast destroyed
4	131–155 mph (210–249 kmph)	Failure of nonbearing exterior walls and some roofs; major damage to windows and doors; larger buildings near the coast suffer flood damage to lower floors
5	greater than 155 mph (249 kmph)	Many roofs and some buildings completely destroyed; major flood damage to all low-lying buildings near the coast

Source: National Hurricane Center website www.nhc.noaa.gov/aboutsshs.shtml

Finally, as the hurricane nears land, Doppler radar is used for more exact measurements of the size and swath of the storm. Winds from hurricanes can extend as far as 171 miles (275 km) in front of the storm's edge.

Once a hurricane moves over land, it usually marks the

Gale force winds lash the seashore and rip palm leaves from the tops of mature plants. Palms are well adapted to hurricanes and although they may lose their leaves, they are rarely uprooted.

beginning of the end of the eye of the storm because it is cut off from the fuel of warm ocean water. If, however, the storm crosses the Florida panhandle and moves back onto the warm water of the Gulf of Mexico, it will refuel and continue to be a hurricane. In 2004, Frances created not only massive flooding along the eastern seaboard of the U.S. but spurred 92 tornadoes in its wake.

Hurricanes are powered by the sun's energy as it is absorbed in the surface layer of the ocean and subsequently transferred to the atmosphere by evaporation and condensation of water. As ocean temperatures rise, the hurricane region will expand into more northern parts of the eastern seaboard,

Many coastal homes along the Gulf of Mexico and elsewhere are built on stilts six to nine feet (1.8–2.7 m) above the ground. When flooding occurs, these dwellings have a greater chance of surviving. The home on the left was not built on stilts and will at the very least suffer from extreme mold damage once the flood waters recede.

and the hurricane season could be extended. Warmer ocean temperatures could be translated into higher wind velocities and larger storm surges.

As our climate changes, precautions will need to be taken for tens of millions of people who live in the southeastern United States.

Galveston 1900

The twentieth century began with the retreat, for the first time in 35,000 years, of the Bering Glacier, in what was not yet the state of Alaska, spawning rivers, calving ice bergs, and eventually shrinking some 650 feet (198 m). That same summer saw unrelenting high temperatures throughout the southwestern United States. It was a record-breaking month of August in 1900 from the Mississippi and Ohio Valleys to the Mid-Atlantic States.

This was a sanguine time, especially for people in the West and particularly in Galveston, Texas. It was the biggest cotton port in the country and third largest port in the nation. It boasted 45 steamship lines, some journeying as far as Europe. Galveston, with 42,000 people, was very affluent. There were more millionaires per square mile than Newport, Rhode Island. Galveston had electric streetcars, electric lights, local and long distance telephone service, two telegraph companies, and three concert halls.

Galveston also had some of the finest and most beautiful beaches on the globe. As is often the case with the world's most inviting locales, Galveston is perched in a rather precarious natural setting. The city of Galveston occupies the eastern end of Galveston Island, a 27-mile-long (43 km) and 3-mile-wide (5 km) barrier island at the southern end of Galveston Bay along the Gulf of Mexico. At its highest point, the island is about 9 feet (2.7 m) above sea level. That means when the tide rises by one foot (30 cm), 1,000 feet (305 m) of beach is lost around the entire island.

Isaac Cline was in charge of the Weather Office in Galveston. He was a keen, meticulous scientist. The main responsibility of the office was to collect weather data, and using Morse code telegraph, communicate the information to Houston and other weather offices. Forecasting was in its infancy. The basic understanding behind hurricanes was incomplete.

On or around August 25, 1900, an upper-air easterly wave occurred over West Africa and drifted into the Atlantic. Ernest Zebrowski, author of *Perils of a Restless Planet*, wonders: "Could a butterfly in a West African rainforest, by flitting to the left of a tree rather than to the right, possibly set into motion a chain of events that translated into a hurricane striking coastal South Carolina a few weeks later?"

Some 70 or 80 upper-air easterly waves occur each year, yet only a growing handful take flight like Zebrowski's butterfly analogy of nonlinear dynamics and chaos theory. Although only initially small, their effects are manifested in the most fearsome storms on the planet — hurricanes.

On August 27, moving at 7 miles per hour (11 km/hr) below the Tropic of Cancer and halfway between Cape Verde and the Antilles, the awesome storm gained strength. By the next day it had traveled 300 miles (483 km) southeast, picking up energy from the warm ocean by adding trillions of water-vapor molecules and gaining speed. Three days later, on the 30th, it was just off the eastern coast of Antigua. From there, the hurricane gained strength, wreaking havoc and heading southeast toward Cuba. It veered over northeast Cuba and turned left, avoiding Florida and only removing the telegraph line between the keys and mainland.

The hurricane then set its sights on Galveston, Texas, some 800 miles (1,287 km) away. It gained incredible heat energy from the very warm Loop Current, which passes through the Straits of Florida and fuels warm water into the Gulf Stream.

Without satellite information, Isaac Cline and all the other weather scientists of the day didn't know about these

The mighty Mississippi Delta was still swollen with floodwaters weeks after Katrina landed.

types of storms, the winds and the resultant swells they created. In particular, they were unaware of what happens when deep-sea swells enter shallow water. Immense amounts of water pile up very quickly. Incidentally, that's what happens when an earthquake occurs under the ocean. The quake creates a tsunami and barely detectable small bump waves move across the deep sea at speeds in excess of 400 miles per hour (644 km/hr) until they hit the shallow shore's edge. The water then piles up and smacks the land with a tidal wave.

On the afternoon of September 8, 1900, Isaac Cline and the Weather Office had no idea a Category 4 hurricane with winds of a least 135 miles per hour (217 km/hr) had struck Galveston with a storm surge of 15.5 feet (4.7 m). The first indication of the enormity of the disaster that befell Galveston was when Isaac Cline saw a wall of water coming toward him. "I was standing at my front door which was partly open, watching the water which was flowing rapidly from east to west. The water rose 4 feet (1.2 m) in 4 seconds. This was not a wave but the sea itself."

Some of the wind gusted over 200 miles per hour (322 km/hr). That is the equivalent of more than 60,000 pounds (27,216 kg) against a house wall, or 30 tons of pressure! The wind and water packed a one-two punch that annihilated at least 8,000 and perhaps as many as 12,000 people in Galveston. Most were swept out to sea, their bodies never recovered.

It was the storm surge and the wall of water that inflicted the worst damage, claiming the highest number of recorded casualties in the United States. Storm surges result from wind. Gusts of up to 200 miles an hour (322 km/hr) are able to move vast and truly unfathomable amounts of sea water inland. The angle at which the surge strikes land and the topography of the land are also important factors determining the severity of the strike. On September 8, 1900, the wind blew from the north, carrying unobstructed water for about 35 miles (56 km) across Galveston Bay, hitting the city of Galveston straight on. In a matter of seconds, 15.5 feet (4.7 m) of storm surge submerged Galveston.

The hurricane, which was dubbed by some as "Isaac's Storm" after Isaac Cline, made front-page headlines worldwide. A global outpouring of generosity was funneled through the Red Cross. California newspaper titan Randolph Hearst gave $50,000 — the equivalent today of $900 million! New York State raised $93,695.77. Money was received from cities around the world, including Liverpool, England, and Moose Jaw, Canada.

KATRINA 2005

One hundred and five years later, fueled by global warming and steamy Gulf of Mexico water temperatures, another hurricane came ashore and devastated Louisiana, Mississippi, and Alabama.

Let's first examine Katrina, then look at New Orleans, its geography, and the Mississippi Delta, and find out why the vitality of this world-renowned ecosystem is so important and how it was naturally designed to mitigate storm surges.

Category 5 hurricane Katrina was one of the strongest hurricanes ever recorded in the Gulf of Mexico. Eighty percent of New Orleans flooded and remained under water for two weeks. It took that long for 148 pumps sucking 90,000 gallons

HURRICANES 37

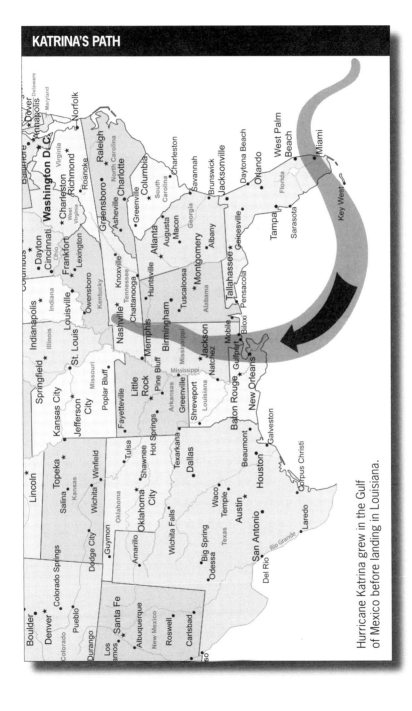

Hurricane Katrina grew in the Gulf of Mexico before landing in Louisiana.

per second (340,687 liters per second) to drain New Orleans.

New Orleans was built on the edge of one of the most fertile and exquisite wetlands in the world — the Mississippi River Delta. The deciduous bald cypress trees (*Taxodium distichum*) live in the swamps of the Bayou with their roots sticking into the air like periscopes. These roots, called pneumatophores, draw air, enabling these trees to live in inundated soils. These marvelous trees are covered with Spanish moss that provides habitat and food for bats, birds, salamanders, and other amphibians. Throughout these swamps, the fearsome alligators, descendants of prehistoric dinosaurs, rule the waterways.

New Orleans itself was constructed on a rare strip of high ground running along the Mississippi River. Its crescent shape gave the city one of its nicknames — The Crescent City. By the mid-19th century, New Orleans, which commanded an enviable trade route up the Mississippi into the heart of America's interior, was one of the wealthiest cities in the country. It had limited land for expansion, but it needed to grow.

Like Holland, it looked to an elaborate system of canals or levees to drain the water from the land and expand. The land was drained and reclaimed, and New Orleans grew into an international port city and magnet for tourism. But when humans work against Mother Nature, no matter how ingenious the scheme, eventually she catches up with us.

First, as the spongy New Orleans reclaimed soil dries out, it shrinks — a process called subsidence.

Second, Louisiana has been pumping oil out of the ground since 1901. The world's first oil platform was constructed in the Gulf of Mexico in 1947. In 2002, a U.S. Geologi-

Mississippi River, MO, July 1993: An aerial view of floodwaters showing the extent of the damage wreaked by the swollen Mississipi River. The river is prone to severe flooding after heavy rain and hurricanes.

cal Survey researcher began to examine the rates of shrinking (subsidence). After 100 years of removing millions of barrels of oil, trillions of cubic feet of natural gas, and tens of millions of barrels of saline water, a drop in subsurface pressure or regional depressurization had resulted. In effect, what's happening is that underground faults are slipping and the land above is slumping.

Third, the Mississippi River periodically floods downstream, carrying fresh silt that replenishes the wetlands over thousands of years, nourishing the plants and filtering the lifeblood of planet Earth — its fresh water. This tremendous process created a large delta of swamps, the Bayou, and barrier

HOW TO PREPARE FOR A HURRICANE WATCH...

1. Track the hurricane progress on a battery-operated radio

2. Top up your car's gas tank

3. Check your emergency supplies

4. Move outdoor items into a garage or shed; anchor any items that can't be stored

5. Close and board up windows

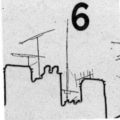

6. Remove outside antennas

7. Turn your refrigerator and freezer to their coldest settings and avoid opening them

8. Fill tubs, jugs, and bottles with drinking water

9. Store valuables in waterproof containers in the highest parts of the house

10. Review your evacuation plan

islands that absorb storm surges and protect inland areas.

The levees that allowed New Orleans to grow destroyed its natural hurricane barriers. With the installation of levees, the silt is funneled directly into the Gulf of Mexico, not spread out naturally over the delta. The drained land is sinking, allowing salt water to penetrate the remaining delta. The salt water is killing plants and trees that are not adapted to saline conditions. It is also destroying habitat for myriad wetland animals.

Every two miles (3.2 km) of wetlands can absorb one foot (30.5 cm) of a storm surge, but Louisiana is currently losing one acre (0.4 ha) of wetlands, roughly three football fields, every 4.5 hours.

When Katrina landed on shore in New Orleans on August 29, 2005, she delivered a one-two-three punch: the storm, the flooding, and the unorganized official response.

The levees gave way in the first day, looting set in on the third day, and dead bodies floated for a week before being taken to the morgue.

In all, the wrath of Katrina spread over 6,400 miles (10,299 km) of the Gulf coast or 140,000 square miles (362,598 km^2). Hundreds of thousand cars and 75,000 boats were demolished in Louisiana alone. A thousand water systems in Louisiana and Mississippi were damaged. Gas and oil platforms were strewn about like matchsticks along the coastline. Five major oil spills and countless chemical spills sullied the delicate coastal ecosystems. After Katrina, the entire infrastructure of the Gulf Coast — electric power, natural-gas lines, water and sewage systems, highways and bridges, and phone lines had to be either restored or rebuilt in their entirety.

The onset of hurricane season is predictable. The fury is not. It's time to begin to work with Mother Nature and not against her. In 2002, the University of New Orleans implemented a $26-million pilot project just south of the Crescent City. Some of the levees were deliberately breached with a series of gated spillways to spread silt back into the dying marshes. The pilot is working, and the marsh is responding, in part because some of the waters contain fertilizer run-off. At least the plants are not dying, and a slice of the coastline, along with its natural hurricane barriers, is beginning to be restored.

CHAPTER 4
TORNADOES

Tornadoes are the most violent storms on Earth. These awesome vacuum-like vortex funnel clouds, often referred to as twisters, are characterized by extreme low-pressure systems. The air pressure during these storms is so low that people can feel light-headed and short of breath because breathing becomes difficult.

The sound of a tornado is deafening and distinctive; it has been likened to the roar of a B-52 bomber, a million bees, or 100 freight trains. Anyone who has ever survived an encounter with a twister will tell you that it's a life-altering experience.

Let's examine tornadoes and discover why the United States is the tornado capital of the world.

The word tornado derives its roots from the Spanish word *tronada* ("thunderstorm"), which comes from the Latin *tronare,* "to thunder." More than mere thunderstorms, tornadoes are violent, visible, rotating columns of air analogous to the trunk of an elephant sucking water and anything else in its way.

Tornadoes are often no wider than 1 mile (1.6 km) with winds of at least 100 miles per hour (161 km/hr) and occasionally in excess of 315 miles per hour (507 km/hr). Funnel clouds will touch down, dissipate, and then reappear.

They usually last for no longer than 10 minutes, yet for every rule there are indeed exceptions in nature. Dubbed "the Super Outbreak" and lasting about 16 hours between April 3 and 4, 1974, 148 tornadoes pummeled 13 states from Illinois, Indiana, and Michigan southward through the Ohio and Tennessee Valleys into Mississippi, Alabama, and Georgia. The Super Outbreak produced more long-track tornadoes than any other tornado outbreak, killing 315 people and injuring 6,142. Xenia, Ohio, was decimated: 34 dead, 1,150 people injured, 300 homes destroyed, a high school demolished, and 2,100 homes badly damaged.

Records of tornadoes extend almost 1,000 years back in history. A tornado struck England on October 17, 1091. It obliterated 600 homes and many churches.

Modern technology, including satellites, computers, and radar have enormously assisted meteorologists to understand weather patterns. In order to comprehend tornadoes, we need to examine some basic principles of weather.

In the northern hemisphere, low-pressure systems rotate counterclockwise while high pressure systems rotate clockwise. The reverse is true in the southern hemisphere. Lows are often linked together, forming troughs, while highs

An ice-cream cone is but one of a number of different shapes that tornadoes can resemble. Tornadoes occur frequently in Tornado Alley and they follow the migrating jet stream north as spring unfolds.

form ridges. Low-pressure systems in the mid-latitudes are referred to as "wave cyclones." "Cyclone" in this context refers to any low pressure that swirls in the atmosphere.

So many twisters have hit Texas, Oklahoma, Kansas, Missouri, Iowa, and Minnesota that this region is called "Tornado Alley." The logical question is, why are there so many occurrences in this region?

Southeasterly winds blowing near the surface from the tepid Gulf of Mexico carry warm, moist air northward across Texas, Oklahoma, Kansas, Missouri, and bordering states. Southwesterly winds from hot and dry northern Mexico, southeastern Arizona, and southern New Mexico also feed into those states. Meanwhile, the cool wet Pacific brings a continuous series of fronts onto the west coast, while all that warm moist and dry air is converging in the south-central United States. The cool Pacific fronts bring precipitation to the Coastal, Cascade, Selkirk, Purcell, and Monashee Mountains, and the west side of the Rocky Mountains, before rising and becoming drier. With the assistance of strong upper-air winds or jet streams, this cold northern dry air is carried south at high altitude to the west, reaching the vast high plains of Tornado Alley.

It is not absolutely necessary for these three air masses to collide in order for tornadoes to form, but when they do, conditions become perfect for the most violent storms on the planet to occur. There is nowhere else on the globe where these three types of air masses converge so frequently. As a result, at least 1,000 tornadoes a year occur here. Tornadoes do occur in Australia, Bangladesh, Canada, England, France, Germany, Japan, and Russia, but not as frequently.

While tornadoes can be born from several types of thunderstorms, it's the "supercells," or long-lived storms, that produce severe weather. Like hurricanes, the weather conditions that make up a supercell start from the formation of clouds.

When warm, humid air rises over ground, water, or mountains, it moves into a lower pressure. This occurs because the weight of the atmosphere above it decreases as it rises. Rising air expands, cooling as it uses up internal energy. Eventually it reaches a point when water droplets condense and form a cloud.

Those big mounds of puffy clouds that look like they'd make a great pillow are called *cumulus*. Their base normally ranges in height from 3,000 to 7,000 feet (914 to 2,133 m), while their tops can attain heights in excess of 60,000 feet (18,288 m). Any moisture at that height will be in the form of ice crystals or snowflakes. The top of the towering cumulus cloud resembles the head of an iron hammer or anvil due to the influence of the strong jet stream. All the ingredients are now in place for the cloud to transform into a thunderstorm or "cumulonimbus" cloud spawning lightning, thunder, heavy rains, occasional hail, and strong updraft and downdraft winds. When this formation has clouds hanging like bags or "mammatocumulus" clouds, then tornadoes are likely to occur.

Inside the superceled thunderstorm clouds are swirls that contain large and smaller swirls. The largest swirls are known as a "mesocyclones." Mesocyclones birth the fearsome twisters.

Supercells mostly move from the southwest to the northeast and the mesocyclone tornadoes occur at the southwest edge of the thunderstorm. The tail or rear edge possesses very strong updraft winds. In front of the

HOW HAILSTONES FORM

The formation of a hailstone is an interesting phenomenon. Its size directly reflects the intensity of the storm and the strength of the updraft. Hail forms when supercooled water, sometimes at −40°F (−40°C) is carried repeatedly to high altitudes, where it freezes. The more successive trips it is carried aloft, the larger the hailstone. Smaller hailstones often merge with larger ones before falling to the ground.

The size of the hailstone determines the speed at which it impacts Earth. For instance, a hailstone with diameter of 0.4 inches (10 mm) moves at 20 miles per hour (32 km/hr) whereas a 3-inch (76 mm) diameter hailstone travels at a speed over 100 miles per hour (161 km/hr). The largest recorded hailstone came from Coffeyville, Kansas, in September 1970: 5.5 inches (140 mm) in diameter, traveling at a speed in excess of 125 miles per hour (200 km/hr)!

All that remained after a loathsome tornado decimated this Midwestern home was one piece of furniture.

twister are ominous wall clouds; beyond the wall clouds with furnace-like downdrafts are hail, lightning, thunder, and torrential rains.

How does the twister of a funnel cloud actually form?

When warm air forms a cap over cooler air, likened by some to a pressure cooker, the superceled mesocyclone thunderstorm releases explosive energy. Wind shears created by changing ground-wind speeds cause swirling warm horizontal winds to meet swirling cold vertical winds, forming a spin or vorticity.

When the horizontal and vertical winds lock together, a depression is formed in the center, just like the depression in the center of a milkshake in a blender. Usually no wider than a mile (1.6 km), the loathsome twister wreaks havoc, behaving erratically before dissipating.

There are a number of different shapes and sizes of twisters. The cone-shaped tornado resembles an ice-cream cone: wide at the top and very narrow at the base. The rope tornado, as the name implies, is rope-like, stretching and eventually dissipating into thin air. Satellite tornadoes result from smaller tornadoes orbiting larger ones. Waterspouts occur over water.

The seasonal strength of the sun plays an important role in the development of tornadoes. In the latter part of the winter, the Gulf coast region and Florida experience the most tornadoes. In early spring, the southern part of Tornado Alley becomes active. By early July, southern Canada and the northern part of the lower 48 states become prone to tornadoes.

Fleets of geosynchronous satellites are constantly monitoring the weather. Radar, which stands for radio detection and ranging, picks up electromagnetic radiation reflected back from raindrops, snowflakes, and hailstones. Doppler radar scans the skies for thunderstorms and is the best form of detecting tornadoes. Doppler radar works in a similar way to the radar guns used by police to check your speed.

If favorable conditions for the formation of supercells

> **RATING TORNADOES**
>
> In the mid-1950s, University of Chicago climatologist Theodore Fujita developed a scale to rate tornado strength. Its six categories range from F0 with winds of 40–72 miles per hour (64–116 km/hr) and associated damages like shallow-rooted trees being uprooted, to F6 with winds of 261–318 miles per hour (420–512 km/hr) with severe damage like healthy mature trees being entirely debarked.

> **STAYING SAFE IN A TORNADO**
>
> If you are in the area of a tornado warning, take the necessary steps to protect yourself:
>
> **GO IMMEDIATELY** to a safe place like a tornado cellar, bomb shelter, or steel-reinforced building.
>
> **IF YOU ARE** outside, find a ditch or low lying area and lie down.
>
> **IF YOU ARE** stranded in a building, get to the center, preferably in the basement, and keep away from windows.
>
> **IF YOU ARE** in a motor vehicle or mobile home, get out.
>
> **DON'T TRY** to outrun the tornado in a car.
>
> **DON'T STOP** under a bridge or underpass because winds can funnel into them.

occur, the National Weather Service and Environment Canada will issue a "watch" covering 25,000 square miles (64,750 km^2) indicating that severe weather is possible in the watch area. Once the storms become severe, a "warning" is issued, signifying imminent danger.

To give you an idea of the strength of these monsters, a tornado in 1927 in Kansas picked up a 5-ton Caterpillar bulldozer and threw it 500 feet (152 m). Another tornado hit a train in Minnesota in 1931 and lifted five of its coaches, each weighing about 70 tons, off the track.

On May 11, 1953, Waco, Texas, was violently assaulted by a twister. Similarly, on July 31, 1987, Edmonton, Alberta, experienced weather described as "reaching biblical proportions: torrential rains, rivers rising, severe hailstones, and twisters."

Waco, Texas

Waco and its metropolitan area in 1953 had a population of about 85,000 people. Located in east central Texas at latitude 31° north, it has an elevation of about 455 feet (139 m) above sea level. Waco is named after the Wichita Native American tribe. The longest river in Texas, the Brazos River (840 miles [1,352 km] long), runs through Waco, eventually emptying into the Gulf of Mexico. The surrounding native vegetation is a mix of prairie grass, longleaf and loblolly pines (*Pinus palustris* and *P. taeda*), oaks (*Quercus* spp.), mesquites (*Prosopis* spp.) and cacti. It has a semi-tropical climate with short cool winters and an average annual precipitation of 32 inches (813 mm). In 1885, the soft drink Dr. Pepper was invented in Waco at Morrison's Old Corner Drug Store.

Waco is located in the heart of the southern Tornado Alley.

In 1911, the Amicable Building, which at the time was touted as the tallest skyscraper in Texas, was built. Twenty-two stories rose 282 feet (86 m) into the sky. When the founder of the Amicable Life Insurance Company, Artemas Roberts, planned the skyscraper, he ordered it built to withstand any kind of punishment, including tornadoes. The 20,000-ton building was built with 3.7 million pounds (1.7 million kg) of steel and 270,000 pounds (122,470 kg) of iron. The building was also equipped with its own generator to provide all its electricity. The Amicable Building was a landmark in Waco.

On May 11, 1953, the mid-afternoon air was still and the humidity oppressive. To the southwest of Waco, the sky had taken on an ominous look. Just before 4 p.m., enormous hailstones hammered cars, homes, and wildlife just southwest of the city. The northeast edge of the superceled mesocyclone began to envelope Waco and its surrounding residents.

At 4:10 p.m. the funnel dipped down, officially becoming a tornado, destroying a home 3 miles (4.8 km) north of Lorena.

At 4:20 p.m., it ripped into Hewitt. Now it was just at the southern limit of Waco. By 4:30 p.m., another twister had touched down along the edge of the University of Baylor's Stadium.

Reports from eyewitnesses in the Amicable Building said that "the day turned into the night; in a couple seconds the sky was black."

The F5 tornado gusting at 285 miles per hour (459 km/hr) entered the business district with a width of at least a couple of city blocks. Pieces of lumber flew past windows, driven horizontally and embedding firmly into exposed steel girders. Telephone poles were knocked down like tenpins in a bowling alley. City trees were entirely debarked.

Chester Stockton, a local painter and tornado survivor, recalled hearing the wind "like a power saw in a wet board." Plate-glass display windows shattered before his eyes as other store windows disintegrated. The intense superceled low pressure caused another local survivor, Ira Baden, to recollect, "I felt light-headed."

Cars were strewn across the streets and flipped upside down; some were crushed, standing vertically nose first.

At about 4:35 p.m. the twister hit the center of the city, causing many people on the streets to run for shelter inside local businesses. Those who fatefully piled into the wood-framed RT Dennis Furniture store never left. The F5 tornado imploded the building in a matter of seconds, killing all 30 people inside.

Twelve-year-old Harvey Horne and his father sought shelter inside Hub's Barber Shop. Harvey remembers his ears popping and having extreme difficulty with breathing.

Edward Cook, a twice-wounded veteran of World War II, likened the path of the twister and the carnage in its wake to what he'd endured overseas in the war.

As it exited the city, the tornado ripped the roof off the East Waco Elementary School and whacked the old suspension bridge, causing it to rock like a cradle, but the bridge held its ground. City Hall was battered but still standing, as was the Roosevelt Hotel. The Dr. Pepper Bottling Plant, now a museum, was badly damaged. Artemas Roberts's foresight to reinforce the Amicable Building with tons of steel and iron proved invaluable as the Waco landmark stood tall and proud when most of the city center and surrounding homes were leveled.

Waco had suffered an unimaginable tragedy — 114 dead, 1,097 injured, 2,000 automobiles demolished or damaged, 850 homes destroyed or partially ruined, 196 buildings flattened or requiring demolition, 376 other buildings declared unsafe, and property damage in excess of $51 million. It was America's 10th deadliest tornado.

Hundreds of rescuers worked throughout the night of May 11–12, 1953. Thirteen hours after the twister departed, they unearthed Lillie Matkin, who miraculously was unharmed other than superficial cuts and deep bruises.

One hour before the first funnel cloud touched down, a research scientist at Texas A&M University observed an echo-shaped comma on his radar screen. He didn't give the echo any attention nor consult with the Weather Office who had posted a watch for the region. Just prior to the F5 ripping into Waco, five echoes were detected on the Weather Office screen.

Edmonton, Alberta

Edmonton is the largest northern Canadian city, and in 1987 it had a population of about 700,000 people. It is located in central Alberta at latitude 53° north with an elevation of 2,192 feet (668 m) above sea level. The Rocky Mountain glacier-fed North Saskatchewan River runs through Edmonton, bringing the city its drinking water, before moving east carrying drinking water to the city of Saskatoon, Saskatchewan, and finally emptying into Lake Winnipeg (the fifth biggest fresh water lake in Canada).

The main native vegetation type are the boreal forest to the north, made up of willow (*Salix* spp.), aspen (*Populus tremuloides*), poplar (*Populus balsamifera*), birch (*Betula* spp.) balsam fir (*Abies balsamea*), white and black spruce (*Picea glauca* and *P. mariana*), Jack pine (*Pinus banksiana*) and tamarack (*Larix laricina*) trees; and to the west, lodgepole pine (*Pinus contorta*) — Alberta's provincial tree, cottonwood (*Populus trichocarpa*), Englemann spruce (*Picea engelmannii*), whitebark pine (*Pinus albicaulis*), and subalpine fir (*Abies lasiocarpa*).

Edmonton has a continental climate with cold winters and warm summers with an average annual precipitation of 18.8 inches (478 mm). It receives an average of 2,013 hours of sunshine a year.

In 1795, the Hudson's Bay Company built a fort on the banks of the North Saskatchewan River. In 1947, crude oil was discovered south of Edmonton and soon after vast deposits were discovered in central and northern Alberta, making it one of North America's oil meccas.

On July 31, 1987, a superceled mesocyclone occurred in

Edmonton. The first of six reported twisters touched down just after 3 p.m. The storm hit the south suburb of Millwoods, then swept northward along 50th Street. Houses were demolished. Trucks and cars were thrown across 23rd Avenue. The second reported tornado occurred in the Industrial Park of Strathcona. Roofs were lifted off complexes, industrial equipment was flipped upside down, and 12 people were killed.

Next, another twister touched down along the Sherwood Park Freeway just as rush-hour traffic commenced. Traffic was unusually heavy at the start of a holiday weekend. The tornado flung cars and trucks into fields, causing several fatalities before moving onto Evergreen Mobile Home Park. It sliced into the mobile homes like a hot knife through butter, killing 15 people and flattening the entire community.

The storm moved 6 miles (9.7 km) south of the city to Beaumont. There it picked up entire barns, splattering hundreds of cows, pigs, and chickens across open fields.

The F4 tornado, with winds gusting as high as 250 miles per hour (402 km/hr) carved a swath of between 330 feet (101 m) and three quarters of a mile (1.2 km) wide and 26 miles (42 km) long, before finally dissipating.

In the three days that followed, torrential rainfall dumped up to one foot (305 mm) of rain, causing extreme flooding in the Smoky, Wapiti, Simonette, and Kakwa Rivers, which rose by as much as 26.6 feet (8 m).

Hail as large as tennis balls pelted an area of 78 miles (126 km) west of Edmonton. There were 50,000 successful hailstone insurance claims for damage to cars and trucks costing in excess of $200 million. Two people in south

Edmonton were knocked unconscious by hailstones. One of the hailstones was the largest recorded in Alberta, weighing more than a half a pound (225 grams)!

Canada's second-worst tornado hammered Edmonton resulting in 27 dead, 600 injured, more than 1,700 people homeless, and property damage in excess of $300 million.

CHAPTER 5
BLIZZARDS

I grew up on the Great Plains grasslands. Snow shovels and snowblowers were mandatory. In late October, before the first snowfall, I would place dozens of tall colored sticks along the edges of our driveway. These sticks let us know where to start digging out after winter storms.

Winters are brutally cold in the center of North America. About half a dozen times each winter the winds howl, snow or more likely ice crystals fall, and temperatures plummet. Sometimes a significant amount of snow falls, and all of the time snow accumulates in drifts, occasionally huge drifts. In a snowstorm, visibility is limited to at most a quarter mile (400 m), and sometimes the wind blows so hard that you cannot see at all. That's called a whiteout.

If you've ever experienced these winter conditions, then you've endured a blizzard.

On March 24, 1870, *The Vindicator* newspaper of Esterville, Iowa, used *blizzard* to describe an intensely strong, cold wind filled with fine snow. The term stuck.

Every winter, at least a dozen blizzards occur in Canada and the United States. Blizzards occur most frequently in the Prairie Provinces, Atlantic Canada, the eastern Arctic, the Upper Mississippi Valley, and the northern Great Plains. In order for a storm to be officially deemed a blizzard, the winds must be gusting at least 35 miles per hour (56 km/hr) with falling barometric pressure for a period of at least three hours. Snowfalls or ice particles are not a requirement for a storm to be considered a blizzard.

Blizzards along the southern Canadian prairies usually result from strong low-pressure systems traveling eastward or northeastward on the lee (or east side) of the Rocky Mountains. The Upper Mississippi Valley (Nebraska, South Dakota, North Dakota, Minnesota, and southern Manitoba) blizzards often result from two different types of low-pressure systems: Alberta lows coming from southern Alberta, and Colorado lows coming from southern Colorado.

Alberta lows move fast as they ride the polar jet stream. They usually bring only a few blizzards each year to the Dakotas, and those blizzards don't last as long as those from the Colorado lows. The Alberta lows bring extremely frigid

> **FEELING THE CHILL**
>
> Gusting winds cause air temperatures to feel colder on exposed skin than they actually are. This effect is called a windchill. When windchill is extreme, any exposed skin will freeze in a matter of 10 seconds or so. Windchill and whiteout conditions claim on average 100 lives a year in Canada — more than hurricanes, tornadoes, floods, extreme heat, and lightning combined.

polar air and unbearable wind chills into the center of the continent. Colorado lows, on the other hand, bring warmer, wetter air up from the Gulf of Mexico. Consequently, they produce more snow than the drier, colder Alberta lows.

The ferocious winds that are associated with blizzards in the center of the North American continent result from strong pressure gradients that develop between high-pressure systems in the Arctic that drift south into Canada and low pressures of either the Alberta or Colorado lows. The counterclockwise flow of either of these two lows causes strong, destructive winds to whip across the Upper Mississippi Valley during blizzards.

Before examining the Children's Blizzard of 1888 and the lethal blizzard of 1947 in western Canada, it is important to understand some of the properties of snow and how animals cope with winter conditions.

> **SASTRUGIS**
>
> Snow particles under blizzard conditions behave like sand particles in the desert. For instance, snow structures called "sastrugis" form compact, rippled sand-dune-like structures. They are common on the Siberian tundra (treeless north country). These wave-like ridges of hard snow run parallel to the wind and occur in great numbers on the Antarctic continent as well as the snowfields of the Arctic. Although they bear some resemblance to sand dunes, they are constantly changing with the direction of the howling winds.

Snow

Winter is a magical season. Like icing on a cake, snow blankets most of Canada and the northern lower 48 for months on end. It's a time to enjoy skiing, snowshoeing, and skating on frozen lakes and ponds.

A snowflake is made up of delicate needles and plates of multifaceted ice crystals. Under a microscope, the flake is one of Mother Nature's most perfect pieces of art. After its creation, however, the snowflake soon begins to undergo change.

Snow that has accumulated on the ground or on a mountainside is called a snowpack. Snowpacks, as all mountaineers know, are constantly changing and are affected by three main processes: time, internal snowpack characteristics, and external weather conditions.

As a part of the dynamics of a snowpack, new snowflakes become rounded, and eventually smaller ones are absorbed into larger ones until they are all roughly the same size. And it should come as no surprise that the destruction of snowflakes proceeds faster as air temperature rises.

The mechanical strength of snow increases because of the bonding of individual ice grains. That's why, for instance, snow can hang off a roof. The Inuit people discovered the strength of snow and are legendary for igloos. They are constructed with wind-packed snow that attains its strength from the bonding that takes place as the destruction of the initial snowflakes proceeds.

As the snowpack continues to deepen beyond 7.9 inches (20 cm), it develops different

---

HOW TO MAKE A QUIN-ZHEE

The Athapaskan people of the northern or boreal forests modified the theme of the igloo to create an easy-to-construct snow house or quin-zhee.

Everyone who works or plays in snow forests or the alpine (above the treeline) should know how to make a quin-zhee. In an emergency, it can save your life by preventing hypothermia.

To make a quin-zhee, all you need is a snowshoe. It takes about 30 minutes to make a large pile of snow. Leave it to set for about an hour. Do not pack it. The domed outside of the shelter becomes self-supporting because of the increased bonding strength between the ice crystals. Hollow out the inside. You now have a temporary shelter that can be 77°F (25°C) warmer than outside air temperatures.

temperatures at different depths. Typically, the top of the pack is colder because it is influenced by the winter air. The bottom, on the other hand, is influenced by the ground, providing a continuous source of heat during the winter. Ice is converted directly to water vapor (a process called sublimation), and the water vapor rises up toward the top of the snowpack.

For those who enjoy the outdoors in winter, the mature snowpack can present serious danger in the form of avalanches. As the ice crystals at the bottom of the snowpack get smaller and smaller, the layer next to the ground becomes free of ice or hollow. This provides crucial habitat for small mammals like mice and voles, which cannot hibernate, to be able to move around and continuously look for food like seeds. It's this hollow layer, however, that contributes directly to avalanche dangers in the rugged Alaskan, British Columbian, Cascade, Sierra Nevada, and Rocky Mountains.

Sunburn and snowblindness are also concerns for skiers, snowshoers, and other winter enthusiasts. Fresh snow is highly reflective, turning back between 80 and 95 percent of sunlight. It is important to wear sunscreen and sunglasses when outside in the mountains during the winter months.

Snow's heat-absorbing properties can create hidden hazards for outdoor adventurers. As tree trunks absorb the sun's short-wave radiation, they re-radiate heat (as long-wave radiation), and this energy is absorbed with near perfect efficiency by the snowpack. The result is increased melting (sublimation) around the tree trunk, which creates snow wells. Backcountry enthusiasts know to keep at least six feet (1.8 m) away from big tree trunks. Snow near the trunk can cave in, creating a well, and flip you upside down, potentially twisting your knees or ankles.

Wintertime and Mother Nature's frozen wonderland is a good time to get outside and further explore the great outdoors. Understanding snowflake dynamics is crucial in northern climates where it can be a matter of life or death.

Coping with Winter Conditions

As the first snow blankets the land, most of us retreat inside, spending much less time outdoors compared to the previous six months. Most people are well protected as long as they are inside during a blizzard. So how does the rest of the animal kingdom contend with winter and blizzard-like conditions?

There are three basic strategies for surviving winter's rigors: migration, hibernation and resistance. From an evolutionary stance, they amount to choices between avoidance and confrontation. Animals either leave the frozen landscape for warmer and more hospitable lands to the south or, with remarkable stamina and adaptations, they stick around.

At first glance, migration appears to be a safe alternative. It is for some, but like all choices, there are consequences. A 600-mile (965 km) flight costs birds 50 percent of their total body weight as fat. And adding more weight for flight has practical limitations.

In addition, waterfowl have added stresses imposed by humans as hunters gather along migratory routes. Consider that a Canadian goose that leaves Atlin, British Columbia, (latitude 60° north) in early September and arrives near the Salton Sea in southern California (latitude 33° north) may face three months of being a target for shooters. The goose season opens progressively later southward along the route.

A blizzard with whiteout conditions makes driving extremely hazardous, especially if you are caught on a highway. Winter drivers are reminded to carry warm blankets in their trunks and a candle and matches in their glove box. The heat from a burning candle can prevent a stranded driver from succumbing to hypothermia.

Snow is removed from roads and highways with the aid of massive equipment that collects the snow and blows it off the thoroughfare.

Some birds, like herons, are specialists, and they cannot change when environmental conditions change. The slightest cover of ice prevents the heron from making its living: fishing. Others, like fly catchers, need insects and when it gets cold, their food source metamorphoses, so they must fly south. Many birds have no other choice but to migrate.

It would cost a mammal of the same weight about 10 times more energy to run a given distance compared to flying it. Swimming appears to be even more energetically cost efficient than flying. Some marine mammals such as blue whales migrate vast distances (5,000 miles or 8,046 km).

Hibernation is a viable alternative to migration for some.

Hibernating mammals posses a unique ability to lower both their metabolism and body temperature. They cannot, however, drop below 32°F (0°C). If they near the freezing mark then they awaken, restore body temperature to its normal state before re-entering hibernation. That costs a large amount of their fat reserves and could jeopardize their chance of surviving the winter.

Cold-blooded animals like reptiles and amphibians cannot control their body temperatures, so hibernation is their only means of survival. Most seek out safe winter hiding places usually under decaying logs, deep protected rock crevices, or in the burrows of other animals, where they can escape subfreezing temperatures. For instance, the northernmost population of rattlesnakes in the Kamloops, British Columbia, region congregate in rock dens as do the 50,000 red-sided garter snakes at Narcisse Snake Pits Wilderness Park in Manitoba.

For many organisms that do not migrate or hibernate, the only strategy is to resist the cold. Overwintering insects and tree buds, for example, do not generate any appreciable heat, so they must depend upon complex biochemical mechanisms to protect them from freezing temperatures. Mountain pine bark beetles produce a sweet alcohol called glycerol, which prevents ice formation in their body tissues, permitting them to cool, without freezing, to −38.2°F (−39°C).

Many animals are well adapted to contend with snow. Spruce grouse grow extra scales or spines on their feet, called pectinations, enabling them to grip the snow while feeding on icy branches. Caribou, with feet seemingly way out of proportion to their body size, float over snow as if walking on snowshoes. Lynx, too, have huge snowshoe-like paws, which enable them to walk on snow without sinking. And the musculature of moose's legs enables them to move through 5 feet (1.5 m) of snow with ease.

Although the animals of the northern United States and Canada are miraculously adapted to cope with winter conditions, occasionally the severity and longevity of a blizzard, particularly if it occurs toward the end of February and March, will be too harsh, resulting in death. The blizzard of 1947 in western Canada, for example, took a massive toll on the game birds of Alberta, Saskatchewan, and Manitoba.

The Children's Blizzard of 1888

Sixteen million people immigrated to the U.S. from 1850 to 1890. Many of them never left the east-coast cities they arrived at. Only the hardy and brave decided to go west to the Upper Mississippi Valley or the great prairie grasslands. Between 1870 and 1900, about 220 million acres (89 million ha) were stripped of sod, ploughed, and planted. Tens of millions of cows were grazed on the short-grass western prairie for a burgeoning beef industry.

In 1888, the National Weather Service was part of the United States Army and headed by General Albert Myer. Forecasting was truly in its infancy.

The last few days of the year 1887 were bitterly cold along the 60th parallel, which intersects with the northern borders of British Columbia, Alberta, Saskatchewan, and Manitoba. On January 3, 1888 the mercury at Fort Simpson, Northwest Territories, read −37°F (−38.3°C) and falling. An enormous cold air mass began to push south, and by January

8 it reached Medicine Hat, Alberta, roughly 70 miles (113 km) from the Montana border and as far east as Qu'Appelle, Saskatchewan, due north of the Montana/North Dakota line.

Frigid arctic air masses are carried south in the winter by the polar jet stream. Polar jet streams are about 6 miles (9.7 km) above Earth and travel from west to east. Their flow is likened to a massive upper-air river. Speeds of 75 miles per hour (121 km/hr) are average, and the polar jet stream can easily attain speeds of 200 miles per hour (322 km/hr), carrying giant blobs of freezing air over America's northern tier of states. There are usually three to five of these polar jet streams moving and revolving around the world at any given moment in time. These waves, like all waves, move sinuously, or like the movement of a snake. When the wave peaks, that's where a "ridge of high pressure" sits, and in its valleys is where the "trough of low pressure" exists.

During the first week of January in 1888, high over the Arctic, whirlpools of polar air about 600 miles (965 km) wide called cold-core lows drifted south. Concurrently, a sudden shot of energy, perhaps from the deserts of Mongolia, surged into a segment of the jet stream, analogous to a high-speed train, called a jet maximum or jet streak. As the cold-core low drifted south, it collided and unraveled its energy into the potent jet streak.

On January 10, the crest of that jet streak reached northern British Columbia or the southern region of the Yukon, where it encountered the immense wall of the northern Rocky Mountains. As that flow descended onto the eastern face of the Rockies, the air warmed slightly. At the same time, a low pressure system to its south sent a massive vortex spinning counterclockwise to the surface.

The roiling flow of the northern jet streak began to push the system southeast. The air was so cold it had a limited capacity to hold moisture; hence, not much snow fell. As the fast-moving low pressure moved southward, it began to crank itself up into a powerful "mid-latitude cyclone."

On January 11, it crossed the Alberta-Montana border. A pool of Arctic air in central Alberta settled in, and the temperatures plummeted. To the south, a mass of unseasonably mild and humid air from the Gulf of Mexico was pushing up over Texas into Okalahoma. The energy between these two distinctly different air masses was huge. In fact, as they came together, violent weather resulted.

The early morning of January 12, 1888, was unseasonably warm along the Dakota Territory and Nebraska plains. The warm winds were welcomed by the hardy prairie dwellers. It was so warm that the school children didn't wear jackets, hats, or mittens to school.

What the Weather Office didn't know then was that cold fronts, jolted by jet streaks, could move so quickly across the prairies that standing water ices up in ridges, small animals instantly freeze to death, and wet clothes on people become immediately encased by ice.

As the cold front descended upon the Dakotas, as many as a billion water droplets were packed into every cubic yard inside the mushrooming clouds. At 6 a.m. on January 12, temperatures in North Platte, Nebraska were 28°F (−2°C), some 30°F (17°C) warmer than Helena, Montana — 670 miles (1078 km) to the northwest.

The Weather Office in Bismarck, North Dakota, was the first to receive a telegraph informing the attendant that temperatures were falling rapidly and gale-force winds were reported in Montana. Just as the report came in, he saw children heading to school. He told them to go home: "The worst blizzard of the season will be here in two hours." Most of the children went home; those who didn't, perished.

The blizzard hit with such ferocity that any person or animal in its wake stood no chance of survival. Temperatures dropped 18°F (10°C) in minutes; 70-mile-per-hour (113 km/hr) winds brought heavy snowfalls with zero visibility.

At 10:30 a.m., the storm hit Groton School, in Groton, North Dakota. The teacher decided to dismiss school immediately. Within minutes, anyone outside the single-roomed schoolhouse succumbed to hypothermia.

By 1 p.m., the blizzard covered most of the Dakotas, two-thirds of western Nebraska, and the northwest of Minnesota.

In Ord, Nebraska, Minnie Freeman's schoolhouse, made of sod, was hit so hard that the tar and sod roof was ripped right off. Somehow she managed to get all of her pupils together and very quickly and successfully got them out of the sod house. They crawled a quarter of a mile (400 m) to the nearest house. Minnie Freeman and all the students lived.

Stella Badger in Seward, Nebraska, 100 miles (161 km) southeast of Ord, failed to act decisively. She did not dismiss her one-roomed class. They all froze to death in the classroom.

The blizzard was so intense that the air tingled with electricity — a phenomenon sailors refer to as St. Elmo's fire. People inside their homes described their hair rising off their scalps, and sparks leaping off the ends of iron pokers as they held them to their stoves.

Temperatures dropped 50°F (28°C) in Helena, Montana,

32°F (18°C) at North Plate, Nebraska, and 55°F (31°C) in Keokuk, Iowa, within hours. This was a tragic blizzard.

Perhaps as many as 500 people died from the blizzard. The killer storm was dubbed the "Children's Blizzard," for most of the fatalities on January 12, 1888, were schoolchildren, ill-prepared because of unseasonably warm morning temperatures for the deep-freeze blizzard they were thrust into.

A few days later, the *New York Tribune* reported that many of the dead were found with torn or missing clothes. Just before the body's core temperature sinks below 88°F (31°C), victims gasp for fresh air. Medical doctors term this "paradoxical undressing."

The Great Western Canadian Blizzard of 1947

The third week in January 1947 was unusually cold across western Canada, the Yukon, and the Northwest Territories. Several blizzards raged across western Canada, and in particular south Saskatchewan was slammed with heavy snowfalls.

Perhaps the hardest hit of all utilities was the Canadian Pacific Railway. Its main route from the prairies through the Rockies and Roger's Pass (4,534 feet or 1,382 meters) down the mighty Fraser Valley and onto the port of Vancouver was a crucial run for the commodity of wheat and grains that supplied 45 million people in England their daily staple foods. A series of blizzards that raged for 10 days cut off this western Canadian pipeline, and the disruption was greatly felt across the Atlantic, where grain supplies were desperately low.

On January 30, the city of Calgary, Alberta, was paralyzed by heavy snowfalls, blowing winds, and bone-chilling temperatures. The city's streetcar system was forced to a standstill. Polar air blanketed the rest of the nation. The forecast for southern Alberta was more snow and extreme temperatures of −40°F (−40°C).

The polar jet stream also brought unseasonably frigid air to southern Britain. Temperatures were well below freezing, heavy snow fell in Kent, and the Thames froze at Windsor. There were fuel shortages as well as gasoline and electricity rationing. Fishing fleets were stormbound and all the roads throughout southern Britain were ice-covered.

By January 31, the southern Albertan blizzard had spread across into southern Saskatchewan with lethal nighttime temperatures of −40°F (−40°C). Much farther north, along the Alaska-Yukon border, the temperatures sank to −78°F (−61°C).

By February 3, Calgary had begun to dig itself out. On the other hand, conditions worsened dramatically for southern Saskatchewan and Manitoba. Transportation was crippled or at best at a standstill. In Saskatchewan, some communities were completely cut off and running out of fuel and food. Railroad officials declared snow conditions were the worst in Canadian railway history.

In Regina, the capital city of Saskatchewan, the two largest department stores and the two largest bread stores were closed. In Winnipeg, the capital city of Manitoba, the electric company canceled streetcar service.

By February 4, a third person had died of hypothermia in Saskatchewan and many communities were desperately in need of fuel and food. Meanwhile in Winnipeg, 193 men worked teams of horses around the clock in bitterly cold conditions to clear the thoroughfares.

Thousands of cattle died from freezing temperatures and deadly windchills across Saskatchewan and Manitoba. Another strong low-pressure system swept over Calgary, dumping 1 foot (30.5 cm) of fresh snow. And snow from that system fell as far south as Sioux Falls, South Dakota, where rescuers worked around the clock to free 200 passengers stranded in automobiles from the snowfall the night before.

On February 7, the mayor of Regina issued a frantic call for coal. Yet another blizzard with 50-mile-per-hour (80 km/hr) winds blasted western Canada.

The Manitoba Federation of Game and Fish called on schoolchildren, farmers, and any outdoor enthusiasts to put feed out for ring-necked pheasants, Hungarian partridges, and sharp-tailed and ruffled grouse. Heavy snowfalls covered natural feed for the birds, and reports of high mortality across the province were widespread.

By February 8, aircraft were preparing to fly food and fuel to starving and freezing towns across Saskatchewan.

The blizzards finally abated on February 10, 1947. Ten successive days and nights of relentless wind, snow, and freezing temperatures was a Canadian record, which still stands today. There are some other very impressive records that were also set during this spate of wild weather. On February 3, the coldest North American temperature of −81.4°F (−63°C) was recorded at Snag Aerodrome, Yukon, a record that has yet to be broken. One train in Saskatchewan was buried in a snowdrift 3,280 feet (1 km) long and 26 feet (8 m) deep. Most farmers across western Canada had to cut holes in roofs of their barns to get at the animals inside. Some railway lines remained impassable until spring melt.

CHAPTER 6
ICE STORMS

Ice storms are one of nature's most treacherous wild weather events. They cause death to humans, animals, and plants and destruction to buildings, utility poles, and power lines.

Ice storms result from warm, moist, low-pressure systems moving over ground-level air masses at or near freezing temperatures. As the super-cooled water droplets come in contact with exposed objects at ground level, they instantaneously freeze. This is sometimes called glaze or rime ice.

At first glance, the frozen winter wonderland looks rather picturesque. In reality, everything is extremely slippery. Roads and sidewalks are sheets of ice, just like a big skating rink. After three-eights of an inch (10 mm) of ice accumulates on power or telephone lines, they begin to snap because of heavy loadings. Tree branches and trunks also buckle and break from excessive weight.

Usually, ice storms occur a few hours at a time intermittently over a period of between 45 and 65 hours before the warmer air slowly displaces the cold air, the ground temperature rises, and the ice melts. If, however, a strong wind ensues before melting occurs, then the damages from the weight of accumulated ice and falling objects can be horrendous. Storms that last longer than 24 hours result in devastating losses, with destruction that runs into the billions of dollars with loss of millions of trees, tens of thousands of animals, and many human fatalities.

Ice storms occur in Europe, Canada, and the United States. They are common from Ontario to Newfoundland, and in the United States they occur in the south central Great Plains, across the Ohio River Valley, and into the mid-Atlantic and New England regions. Ice storms also take place in mountains. Another prone region is the Columbia River Valley between Oregon and Washington state. Warm Pacific storms blanket cold air trapped in valleys, resulting in fierce ice storms.

In nature, the role of ice storms can be viewed as creating an opening or a "disturbance" in the forest. The frequency of ice storms in Canada and the United States ranges between 20 and 100 years. Branches are often snapped in addition to trunks, but trees are rarely uprooted. The initial damage to the trees creates opportunities for airborne fungus and bacteria to begin to break down or decompose wood. This attracts insects, which draw insect-feeding birds and cavity dwellers, including raccoons, to the trees.

Initial ice damage creates a lack of food for ground-dwelling game birds like grouse, partridges, and turkey and some smaller ground-dwelling animals. Many studies clearly show that the forest quickly recovers. Ice storms create openings that allow sunlight to reach the forest floor and promote natural tree-seedling regeneration. Incoming sunlight enables seedling leaves to make food, build their woody structures, and race toward the sun. Disturbances promote a diversity of tree species with a range of ages.

Several recent studies have shown that plant and animal richness and diversity in Quebec forests has significantly increased four years after the 1998 ice storm in that province. Furthermore, some forestry experts recommend that forest management emulate this method of creating small openings rather than large clear-cut harvests in eastern hardwood forests.

Interestingly, downed branches and trunks from ice storms play a crucial role along streams and lakes. Large amounts of wood in these waters increase the nutrient retention and enrich fish habitat.

Though ice storms can create havoc for humans by causing power outages and making transportation impossible, plants and animals have evolved to contend with ice.

Ice and Plants

Plants cope with snow and cold winter temperatures with a variety of different strategies. Unlike some animals, they cannot migrate to warmer environs.

Many herbaceous plants do the next best thing to migrating: they shed all of their above-ground parts and spend winter safely beneath the soil. Aspens, maples (*Acer* spp.), birches, alders (*Alnus* spp.) and other deciduous trees protect themselves from winter temperatures by dropping their leaves. Native conifers, except for larches (*Larix* spp.), are evergreen and so they have internal functions or physiological adaptations that help them get through the winter months ahead.

The two most common stresses among trees and shrubs of the north country are low temperatures and drying up, or desiccation.

In order to prepare for winter, leaves of northern-hemisphere plants begin to recognize the diminishing length of daylight in August. Certain plant hormones are released to slow and then eventually stop all growth. The first frost of the autumn prepares woody plants for the impending onslaught of winter. In addition, plants experience a water stress, which further prepares them for the chilly months ahead.

Trees are then able to deal with freezing temperatures and the controlled formation of ice. The exact location of ice within the tree is very important. Most of the cells within trees are non-living, because their role is to conduct water during the growing season and provide mechanical support or stability. There are, however, living cells within the branches, trunk, and evergreen needles that are very important for storing food and kick-starting spring growth. It's within these cells that the exact formation of ice is a life-or-death matter.

The initial formation of ice occurs outside the living plant cell in a small space within the cell wall. All the water that isn't bonded to other molecules inside the cell is exported to the space in the cell wall. When ice forms in the cell wall, it attracts water to its crystals. The living part of the cell is protected by an elastic cell membrane and the remaining cell sap can withstand temperatures as low as −48°F (−55°C). If, for any reason, the cell membrane becomes ruptured or if too much water is exported into the cell wall, the cell sap will become toxic and the cell will die.

Exposed evergreen needles face the greatest water-loss problems under bright sunshine and calm winter days. The needles are warmed to above-freezing temperatures and the air is dry, creating atmospheric suction or a call for water from its needles. The tree is faced with a problem: it's losing water in its winterized needles and must replace it.

The trunk, being darker and warming above freezing like the needles, is able to supply minimal amounts of stored water from the cell walls. This becomes a tricky balancing act. On a day such as this, evergreen trees prefer even the slightest breeze, because that cools the needle surface and prevents any moisture loss and subsequent demand for replacement water.

Heavy snow loads, particularly on the Alaskan, Sierra Nevada, Coastal, or Cascade Mountains, can cause entire trees to bend. A 40-foot (12 m) tall Pacific silver fir (*Abies amabilis*) can accumulate a mass of snow and ice nearly 20 inches (51 cm) thick, weighing 3.3 tons.

Exposed areas of trees are subjected to blowing ice, which can remove foliage or cause freezing injury and create deep pits, eventually wearing away tree bark. Mountain winds, especially during the winter, shape trees and the treeline forests. Some high elevation trees actually resemble broomsticks with windswept branches and trunks with only a "mop-head" or cluster of foliage at their top.

Browsing activities of mammals create further winter-stress problems for plants.

Yet despite all the harsh winter environmental conditions, hardy trees can live for hundreds and sometimes thousands of years, toughly facing months of winter.

Ice and Animals

Surviving winter in the great outdoors is difficult for animals, especially those whose body temperatures are unregulated and subjected to subfreezing temperatures. Insects and amphibians have some remarkable adaptations that enable them to successfully overwinter.

The dangers of ice formation in insects and amphibians are similar to those of freezing in plants. Ice must be prevented from growing within cells, instead growing in a controlled fashion in spaces between cells.

Insects either allow themselves to freeze, or they avoid freezing altogether. Neither strategy is without risks.

Insects that avoid freezing, like mountain pine bark beetles, actively lower their super-cooling point and prevent ice formation by lowering their freezing point, enabling them to tolerate temperatures as low as −38.3°F (−39°C).

In order to achieve this level of tolerance, they select a dry hibernating site. An insect's waxy outer layer is impermeable to water, and it serves as an excellent mechanical barrier against external ice seeding. Since one organ within bugs that is particularly susceptible to freezing is the gut, insects empty it, completely, before the onset of hibernation.

In order to super-cool their blood (correctly called hemolymph), non-freezing insects manufacture antifreeze proteins, alcohols, sugars, and even ethylene glycol (the same compound we use in our car radiators). The correct mix of anti-freeze compounds blocks ice formation and prevents its growth.

Supper-cooling is a gamble. At any temperature below 32°F (0°C), the only stable state in which water can exist is solid. Any kind of disturbance for a super-cooled bug results in spontaneous flash freezing — an instant death.

The only way to avoid this risk is to freeze. Some organisms promote early and gradual freezing of fluids between cells, more specifically in the cell membrane. This controlled freezing is a strategy used by a variety of insects, some frogs, garter snakes, and hatchlings of the painted turtle.

There are four requirements in order to successfully freeze: ice growth is promoted at temperatures just beneath 32°F (0°C); it must be restricted to spaces between cells; the total amount of body ice is limited; and cell membranes must be protected from structural damage.

The ability to create early ice formation is unique, and it is undertaken by special ice-attracting proteins in fluids between cells in their cell membrane. These proteins are manufactured in the autumn and disappear in the spring. They essentially reduce energy barriers needed to make ice and promote a lattice (just like a fence) for ice formation.

Here's the tricky part: If ice grows unchecked (i.e., larger and larger), it will dehydrate the inside of cells. The lethal limit of ice, in a living organism, seems to be about 65 percent of total body weight. In order to control ice, once its formation is induced, freeze-tolerant organisms must also make anti-freeze proteins to keep ice masses small.

Some bugs in the far north are so well adapted that they are able to switch yearly between freezing or super-cooling!

How do insects know when to get ready for winter? Just like plants in the northern hemisphere, insects rely on low, above freezing, air temperatures and reduced daylight hours as their cue to get ready to freeze or get on the super-cooling roller coaster.

Land-hibernating frogs, with water permeable skin, must withstand freezing. They overwinter beneath a scant cover of leaf litter under snow where they may experience subfreezing temperatures and are not able to control their body temperatures. In one of Mother Nature's most awesome winter feats, spring peepers, chorus frogs, gray tree frogs, and wood frogs freeze solid, resembling hard baseballs.

They do this by initiating freezing at 28°F (–2°C) with high amounts of the sugar glucose (rather than alcohols, which bugs use). Glucose is extremely important in protecting membranes like skin.

Very suddenly, glycogen (stored glucose) in the liver is converted to glucose (sugar) and dumped into the blood stream. The glucose level increases 200-fold in about eight hours, until the frog becomes severely diabetic.

Not only is this risky, but it becomes a race to deliver the sugar to the body tissues while freezing is progressing. A strong heart is mandatory, and within one minute of the first ice formation a frog's heartbeat doubles.

Within 20 hours after initial freezing, 60 to 65 percent of the frog's body water is frozen. Its heart stops, and breathing ceases. The frog teeters on the edge of life. The thin litter layer and a good insulative snow cover are of paramount importance. No frog has yet been found to survive below 19°F (–7°C).

Once temperatures rise in the spring, within an hour of thawing, the frog's heart resumes beating. Six hours later, when the frog's body temperature reaches 41°F (5°C), the heart rate is back to normal.

To freeze or not to freeze? It's a question that Mother Nature has at least two exceptional answers for.

The Ice Storm of 1998

On January 1, 1998, a huge low-pressure system over Texas was sucking moist, warm air from the Gulf of Mexico as far north as southern Ontario and Quebec, some 2,500 miles (4,023 km) away. At the same time, a massive stationary Arctic high-pressure system centered over Hudson Bay created a northeast flow of frigid air over central Quebec, which drained into the St. Lawrence and Ottawa river valleys. This strong high-pressure system sank to ground level.

The weather systems remained unchanged for days because a Bermuda high off the Atlantic coast blocked the Gulf of Mexico storm from moving off North America northward to Iceland. Instead, the Bermuda high pushed the moist Gulf of Mexico low northward along the western edges of the Appalachian Mountains into eastern Ontario, Quebec, northern New York, Vermont, New Hampshire, and Maine, where it met head-on the cold arctic air hugging the ground.

For a period of five days between January 5 and 10, the fiercest ice storm to pelt North America in modern times took place. For more than 80 hours, the moist Gulf of Mexico low sat over the freezing Arctic high, delivering steady freezing rain and drizzle over several thousand square miles.

One particular area east of Montreal, St.-Jean-sur-Richelieu, received a record 4.7 inches (12 cm) of ice. The area was dubbed "the triangle of darkness" after the large steel pylons carrying electricity lines all around St.-Jean-sur-Richelieu fell like dominoes. The region went without electricity, in the middle of winter, for almost a month.

In total, the storm destroyed 74,564 miles (120,000 km) of power lines and telephone cables, 130 major transmission towers, and 30,000 wooden utility poles. Over 1,000 steel pylons carrying electricity collapsed from the weight of ice, leaving more than four million people without electricity in southern Quebec, western New Brunswick, eastern Ontario, upstate New York, and northern New England. Maine governor Angus King called it "the worst power outage we've ever seen."

All the bridges linking Montreal, a city of 2.5 million, to the south shore of the St. Lawrence River were closed because of the weight of ice and ice chunks falling from the steel bridge girders. The island of Montreal was without power for six hours, and all the water-pumping stations were shut down. By January 7, the Canadian Armed Forces were deployed across Ontario, Quebec, and New Brunswick. It was the largest non-war deployment in Canadian history.

Electricity was not restored to some 700,000 people until early February, three weeks after the storm had subsided. The death toll from the ice storm in eastern Canada alone was somewhere between 25 and 44 people. Damages were estimated in excess of $7 billion. Collette Fontaine, who sought refuge at a makeshift shelter in Montreal, succinctly remarked, "It's a disaster."

Steel pylons carrying electrical wires buckled under the weight of the ice during the 1998 Ice Storm. Some people went without electricity for three weeks in the middle of winter in northeast and eastern Canada.

Sugar maple trees were hit the hardest by the 1998 Ice Storm. Excessive weight from ice loading killed tens of thousands of mature sugar maples. The sugar maple industry predicted it would take at least four decades before producers could regain 1997 maple syrup production rates.

The ice storm affected over 17.3 million acres (7 million ha) of forests. Birch trees were the most badly damaged, while some of the eastern hemlocks (*Tsuga canadensis*) were somehow able to sustain up to 5.4 tons of ice loading. Sugar maple (*Acer saccharum*) trees were terribly damaged; consequently, the maple sugar industry in Quebec, Vermont, New Hampshire, Maine, and New York suffered millions of dollars of loss. Some experts predicted that it would take at least 40 years before syrup production returned to pre-storm levels.

Farmers, too, were hit particularly hard by the storm. Dairy and hog farmers went without power for days, resulting in millions of gallons of spoiled milk and tens of thousands of animal deaths. "We have cows that have not been milked for a long time," said Vermont governor Howard Dean.

Much farther south, torrential flooding lambasted northeastern Tennessee, North Carolina, Alabama, and Mississippi.

The ice storm of 1998 brought Canadians and Americans from all walks of life together, helping one another to overcome an awful winter disaster.

The Inland Northwest Ice Storm of 1996

On November 14, 1996, the Hawaiian Islands received record amounts of rainfall. The island of Oahu received 20 inches (51 cm) of rainfall with torrential flooding; the average precipitation for November in that area is only 3 inches (8 cm). Four days later, 11.7 inches (29.7 cm) fell on Port Orford, Oregon, the state record for a 24-hour period.

The moisture from Hawaii was carried to the mainland by upper winds called the "Pineapple Express." These winds blow from the southwest and pump warm, saturated air from near the Hawaiian Islands northeastward into California, western Oregon, and western Washington.

A Pineapple Express develops when the Aleutian (Alaska) low strengthens, then shifts southward with the polar jet stream that dips to the south and east of the low. The polar jet stream meets the subtropical jet stream, pulling warm, moist tropical air from the southwest to the northeast. A cold front forms along the converging edges of the polar and subtropical jet streams. The heaviest precipitation falls just south of the cold front.

The west coast typically experiences one to four Pineapple Expresses a year. It can rain or snow at high elevations during a Pineapple Express. When the Pineapple Express remains stationary for a couple of days, its effects translate into devastation.

On the night of November 18, 1996, a Pineapple Express moved across the Pacific into Oregon and Washington, crossing the Cascade Mountains into inland Washington. At the same time, a cold Arctic air mass centered over northern British Columbia swept down into eastern Washington and northern Idaho, sinking to the ground.

Light snow fell over the city of Spokane, Washington, and by morning 4 inches (10 cm) of snow blanketed the ground. Temperatures were hovering at freezing when the Pineapple Express crossed the Cascade Mountains and began to dump precipitation. All the ingredients were in place for the perfect ice storm: moist, warm air over a cold, sub-freezing layer of air on the ground.

For 12 hours, Spokane and the surrounding area were coated by 1 inch (25 mm) of ice, glaze, and rime ice. Strong winds kicked up, and ice-covered power lines and telephone cables began snapping.

By nightfall on November 19, Spokane County in eastern Washington fell into darkness. For many of the quarter-million residents, electricity did not return until on or after December 3, two weeks later.

One third of the ponderosa (*Pinus ponderosa*) and lodgepole pines — hundreds of thousands of trees — in the forested region died. Five hundred homes and businesses were damaged by falling urban trees, and there were five fatalities. The damages from the ice storm exceeded $23 million.

As the storm moved farther east into Idaho, it dumped 3 feet (91 cm) of snow on Bonners Ferry and coated Kootenia, Clearwater, and Idaho Counties with at least 1 inch (2.5 cm) of ice. Damages were estimated at $6 million.

The storm moved up into Fernie, British Columbia, where it dumped another 3 feet (91 cm) of wet snow before the system finally abated.

CHAPTER 7
DROUGHT

Water is the most common yet most important substance on Earth. Its quantity and availability, more than any single environmental factor, determine the type and amounts of vegetation that occur around the globe. Life cannot exist without water.

When precipitation is reduced by as little as 30 percent, a drought begins to set in. Although initially its effects are not as fearsome as those of hurricanes, tornadoes, blizzards, or ice storms, the long-term outcome from drought is deadly — especially when drought lasts for years. Over the past 5,000 years, megadroughts have brought at least three civilizations to an end. More droughts are predicted to occur in the 21st century because of global warming.

In order to understand drought and its effects, it is of paramount importance to understand the lifeblood of our planet: fresh water.

Fresh water is constantly recycled on our planet. Of the water vapor returned to our atmosphere, 16 percent comes from transpiration (water released into the air by leaves in order to take in carbon dioxide [CO_2] to make food) by land plants and the rest comes from the oceans. At any moment, only a thousandth of one percent of the planet's total water is in the air. That infinitesimal percentage produces thick coastal fogs, awesome thunderheads, and soaking downpours. The recycling of atmospheric vapor is so constant that water is completely replaced every eight days, and the equivalent of all the oceans' water passes through the atmosphere every 3,100 years.

Oddly, 97 percent of the water on the planet is saltwater; two-thirds of the fresh water is locked in ice caps and glaciers. Less than one percent of the total is available fresh water, and much of that is in underground aquifers that are not fully accessible. Astoundingly, just 0.016 percent of the lifeblood of the Earth — fresh water — moves through lakes, rivers, snowpacks, the atmosphere, and all living creatures.

The accumulation of snowfall in the winter, termed snowpack, is of critical importance for supplying fresh drinking water, particularly in the summertime, to about 70 million people in western North America. A slow springtime melt allows the meltwaters to recharge the soil, running off into streams, rivers, and forest reservoirs, called watersheds. The southern Rocky Mountains, with the headwaters of the mighty Colorado River in Rocky Mountain National Park, supplies water to most of the American southwest. Over the past 50 years, there has been a steady decline in the amount

When annual precipitation levels drop below 10 inches (25.4 cm) desert conditions set in. Tens of thousands of acres have been rendered fallow by a decade-long drought gripping parts of the Midwest and southwestern U.S.

of snow received both in the southern Rockies and the Cascade Mountains of the Pacific Northwest. Some areas of the southwest and Midwest are experiencing their 11th consecutive year of drought.

The searing summer heat of 2006 in the United States, coupled with hurricanes in Florida and Mexico's Baja peninsula, shriveled tomato yield by as much as 40 percent nationwide. In 2005, a wholesale 25-pound (11.3 kg) box of tomatoes ranged between $13 and $16; the price in 2006 soared to between $30 and $40. In September 2006, Congress agreed to spend $6 billion to bail out farmers across the American southwest and Midwest suffering from drought. In 2002,

30 states felt the wrath of drought, which cost taxpayers in excess of $10 billion. That same year, Arizona suffered its worst drought in 1,000 years.

Drought is causing problems on the other side of the planet, too. Perth, Australia, the most isolated city of 1.5 million people in the world, is running out of drinking water. Since 1975, the rains have significantly lessened. From 1975 to 1996, Perth received 50 percent less water than it did in the first 75 years of the 20th century. A desalination plant at an exorbitant cost of $350 million is scheduled to begin construction, but it will only supply 15 percent of the city's freshwater needs.

On Australia's east coast, the thriving international megalopolis of Sydney, with its population of 4.5 million, is also perilously in need of drinking water. Sydney has about two years of stored water supply left. It, too, has suffered from less precipitation and a prolonged drought that has gripped the entire nation for much of the last five years. Currently, nine sewage treatment plants are pumping recycled waste water into Sydney's drinking water supply. In addition, Sydney may also need to resort to large-scale desalination plants.

In response to the deepening drought in Australia, Prime Minister John Howard's government has created a national office of water resources allocating $1.5 billion to a National Water Initiative in a desperate attempt to secure more water for the coming decades.

In April 2006, Australia's prolonged drought, the worst in at least 100 years, caused the price of vegetables to rise by 8 percent, and the wheat crop was predicted to fall by at least 65 percent from the previous year's harvest of 23.6 million tons.

Predictions for the 2006 barley crop were grim. Australia harvested 11.2 million tons of barley in 2005. A meager harvest of 2 million tons of barley was forecast for 2006. In December 2006, the price of vegetables looked set to rise by at least 30 percent.

West Australian wine growers are concerned that legendary wines from Margaret River may disappear within the next decade or so because of extreme and prolonged drought conditions. "There's going to be a lot of wailing and gnashing of teeth when one concedes you're losing a reputation for world-class wines," said Dr. Richard Smart, an Australian viticulturist.

What Trees Tell Us about Drought

Trees are living museums. Their rings are able to shed light on past climates. Dendochronologists are scientists who study tree rings. They use a non-destructive method of sampling whereby they carefully extract pencil-width cores of wood from living trees. Each year, trees in the West produce a light- and a dark-colored ring of growth. The width of the tree ring reveals growth patterns, which in turn reflect temperature and precipitation patterns, including snowfall. If the distance between rings is wide, that translates into ample moisture, hence good growth. On the other hand, narrow widths are equated with dry conditions yielding little or no growth. Tree rings also exhibit fire scars, which enable scientists to determine patterns and frequencies of lightning-induced fires. Incidentally, counting either the light or the dark rings, but not both, will give an accurate estimate of the tree's age.

Dendrochronolgy is an advanced science, and the

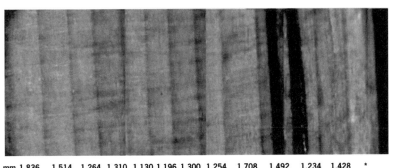

This image shows tree ring growth from a European spruce tree during the Little Ice Age in the 16th century. As temperatures decreased so too did the amount of wood that trees in Europe were able to produce. Consequently, the wood grown during this cold period had few but dense cells. Some of this dense wood was used to make exquisite musical instruments.

Great Basin bristlecone pines (*Pinus longaeva*) of the White Mountains in east central California have given scientists a continuous glimpse of climates during the last 8,700 years. The oldest living tree, "Methuselah," is almost 5,000 years old. It has witnessed almost 1.8 million sunrises. Dead standing and fallen bristlecone pines provide moisture records to 6700 B.C.

Tree rings from bald cypress trees in the Carolinas revealed why the early settlers at Roanoke in 1587 A.D. perished. They arrived at the beginning of a prolonged drought. Bald cypress rings from two decades later showed that the Jamestown colonists arrived during an intense six-year drought lasting from 1606 to 1612 A.D.

There have been some horrific droughts in recent memory. In 2003, there were extreme heat waves and droughts that killed 35,000 people in Europe. In 1988, there were droughts

Droughts not only decimate life on land, but as fresh and salt waters recede, aquatic life suffers, too.

and monster fires in the United States. And in 1972, a drought ravaged crops in the former USSR, throwing the world grain markets in a tailspin. In Africa, along the southern edge of the Sahara Desert in the Sahel region, it failed to rain in 1972 and 1973, resulting in the death of 4 million cattle and 200,000 people.

The Dust Bowl of the 1930s was particularly extreme during the summers of 1934 and 1936 and the autumn of 1936. Fifty million acres (20.2 million ha) of the Plains and Midwest from Texas to Canada and from Colorado to Illinois were affected. Wind storms blew topsoil into roiling clouds of dust.

Have there been other long lasting or megadroughts in the history of civilization? Yes.

The Akkadian Empire

"Through history it has been changes in precipitation, not temperatures, that have brought down civilization," according to Dr. Richard Alley, author of *Two Mile Time Machine*.

The land between the Tigris and Euphrates Rivers, known as Mesopotamia, was a very fertile crescent some 6,500 years ago. Barley was first cultivated there. Hunter-gatherers began to grow crops and graze sheep and goats. They learned how to combine tin with copper to make bronze tools and weapons in the era known as the Bronze Age.

The north of this region, present-day Syria and Iraq, had sufficient rainfall to sustain crops. The southern region was much hotter and drier, and farmers there used elaborate methods of diverting the Euphrates to irrigate the parched lands.

In 2340 B.C., a warrior king named Sargon conquered much of the Middle East and created the first recognized empire, called the Akkadian Empire. Its language was the earliest Semitic language so far identified. Akkadians had a very sophisticated social organization, and controlled the silver mines of Anatolia, the lapis lazuli mines of Badakhshan, and the cedar forests of Lebanon across to the Gulf of Oman. Northern Mesopotamia was the breadbasket where fertile lands grew bountiful wheat.

Archeologists discovered along the Khabur Plain, in northeast Syria between the Tigris and Euphrates, a prosperous outpost named Tell Leilan. Akkad took over Tell Leilan, a city of over 300,000 people, in 2280 B.C.

Akkadian success depended upon feeding its vast empire and hundreds of thousands of laborers, who built schools and storehouses, and ensured that the imperial cities were fortified. They were paid two flat-bottomed bowls of lentils and barley a day.

By all accounts, the Akkadian Empire was quite advanced. Students in one-room schoolhouses learned glyphs and mathematics using an advanced base-60 number system still used today for time and angle measurements.

Goods and personal belongings that were left in Tell Leilan suggest that it was abandoned overnight. The reason for the sudden abandonment and collapse of the Akkadian Empire perplexed archeologists for almost 20 years. Was the entire city overrun and ransacked? Was it a plague? Or was it mass starvation?

The mystery wasn't solved until the mid-1990s, when samples from polar ice cores were examined. The irrefutable cause of the demise of Tell Leilan and the Akkadian Empire was a megadrought that began in 2200 B.C. and lasted until

1900 B.C. — a 300-year span. Polar ice cores at the beginning of the megadrought showed less atmospheric dust and sea salt, meaning that there was less atmospheric circulation, preventing the summertime rains originating in the North Atlantic and Mediterranean — the bulk of the region's annual rainfall — from occurring.

The ice-core data corroborated what archeologists studying soil profiles at Tell Leilan had discovered. They found that the first excavated layer had some earthworm activity, signaling sufficient soil moisture from growing crops, but as the drought set in there was an increased amount of windblown dust and dry soils but no earthworm activity; hence, there was not enough soil moisture for crops. As the aridity intensified, so too did the accumulation of windblown dust. By 1900 B.C., rains began occurring, and people and earthworms returned.

The Mayan Empire

The Mayans of Mesoamerica originated in the Yucatan peninsula of modern-day Mexico around 2500 B.C. They rose to prominence about 250 A.D., occupying southern Mexico, Guatemala, western Honduras, El Salvador, and northern Belize. They developed astronomy, geometric systems, and hieroglyphic writing. They built exquisite ceremonial architecture, pyramids, palaces, and observatories without the aid of any metal tools.

Mayans were expert water managers. And, like the Akkadians on the other side of the globe, they had many high priests. For a period of about 1,200 years, the Mayan culture flourished in spite of an environment with minimal topsoil or water and vicious Caribbean hurricanes.

The Mayans were skilled weavers and potters, and they developed extensive trade networks with distant aboriginal peoples. They were also skilled farmers who were dispersed and somewhat self-sufficient. However, because the seasons were distinctly either dry or wet, water was used as a powerful commodity to attract labor to the Mayan centers.

During the dry months, laborers received drinking and cooking water in exchange for their services. The priests and nobility positioned themselves as the gatekeepers to the fresh drinking water. By controlling the water, they controlled the people. With no lakes or rivers nearby, the people turned to the nobility for water during the dry months.

Water was stored in natural and artificial sites. Some was stored underground in structures called "chultuns," which could supply up to 25 people with a year's worth of water.

The nobility used water lilies (*Nymphaea* spp.) as an indicator that the stored water was potable. Water lilies are very sensitive to changes in water quality, and this ingenious method of determining safe drinking water in a tropical environment enabled the Mayan culture to prosper.

However, in the 800s and early 900s A.D., something occurred that resulted in the demise of millions of Mayans.

Archeologists and anthropologists have put forward many hypotheses, such as internal conflicts, wars, disease, and even famine. It wasn't until the late 1990s that scientists looking specifically at climate, in particular drought, pinpointed the root of the problem.

Lakebed sediments from the Yucatan peninsula clearly showed that three waves of drought descended in cycles of three-, six-, and nine-year intervals, in 810, 860, and finishing in 910 A.D., that wiped the entire Mayan culture off the map. When the farmers arrived at the centers and the priests could not provide water, millions of people dispersed. Some no doubt died through conflicts, others from disease, and most from thirst or famine. It is almost impossible to delineate between death by starvation or disease; they are inexorably intertwined.

Those who believe that history doesn't repeat should take note. About 400 years later, the Anasazi people of the Four Corners region in the southwestern United States — known for domesticating corn, squash, and beans, and building stone houses with elaborate pictographs and petroglyphs — vanished.

Tree ring data from Rocky Mountain Douglas-fir (*Pseudotsuga menziesii* var. *glauca*), ponderosa pine, Rocky Mountain bristlecone pine (*Pinus aristata*), and Engelmann spruce showed that precipitation began to decline in 1276. The climate became significantly drier as the drought deepened until at least 1299.

Drought clearly has the power to decimate even the mightiest of societies.

CHAPTER 8
FIRE

The forests of Australia, Canada, the United States, and elsewhere have evolved with lightning-induced wildfires. Interestingly, it is the frequency and intensity of wildfire that in large part determines the vegetation type. For instance, frequent fires on the prairie grasslands prevent trees from getting established and conquering new territory.

Monster fires are primarily driven by climatic conditions. Furthermore, rising global temperatures increase the intensity and frequency of droughts. And droughts promote large monster-like fires, which release CO_2 into the atmosphere.

Western North America is drying up. The six worst fire seasons in the United States in modern times have occurred in recent years: 1988, 2000, 2002, 2004, 2005, and 2006. That's about 48.1 million acres (19.5 million ha) of forestland consumed in six years, in addition to the 39.1 million acres

(15.8 million ha) burned in the previous 11 years.

The spring and summer of 2004 saw millions of acres ablaze in Alaska, Yukon, and northern British Columbia. The fires were fueled by a prolonged drought at latitude 62° north. Extreme arid conditions in November and December 2005 and January and February 2006 saw millions of acres of grasslands and forests burning in Texas, Okalahoma, Kansas, Arkansas, Nebraska, and New Mexico. Moreover, the winter of 2005–2006 was the warmest on record in New Mexico. In January 2006, climate and forest experts predicted that the upcoming wildfire year would be the worst on record. Thousands of miles north in Alberta and Manitoba, the fire season officially commenced in early April, almost a month earlier than normal.

U.S. fire predictions were correct; the total area burned in the first eight months of 2006 was the highest recorded dating back to 1960, the first year national statistics were recorded. By early October 2006, 9.93 million acres (4 million ha) had burned; that's almost 5 million acres (2 million ha) more than the 10-year average.

On the heals of Australia's worst drought in at least 100 years, the first week in December of 2006 saw enormous firestorms incinerate southeastern Australia, including Tasmania. In the state of Victoria alone, 691,900 acres (280,000 ha) were scorched. On December 9, smoke was so thick from the infernos to the northeast of Melbourne that fire alarms were repeatedly set off at the Melbourne International Airport.

Fire in a wild forest can be thought of as nature's cleansing broom — a change that resets the ecological clock.

Ecosystems in the West depend upon the fire cycle and in fact have developed adaptations to contend with its occurrence. Fire exposes mineral soil and recycles nutrients for seedlings and understory plants, enabling regeneration of burnt-over areas. Rain dissolves the nutrients and returns them to the soil. When fire is suppressed in these western forests, other agents of change, bark beetles, move in.

How Fires Spread

Once a fire is ignited, the rate of spread and its direction are significantly influenced by wind. Furthermore, wind supplies more oxygen, which in turn increases the combustion, essentially preheating and drying fuel in the path of the growing fire.

When a wildfire reaches a critical size, it begins to create its own weather. Enormous rising plumes of heat or cyclonic circulations suck in air from all directions, creating tornado-like winds that can rip a mature tree right out of the ground.

Monster fires spit fireballs as far as half a mile (805 m) in front of themselves, causing ignition of new spot fires. Of more concern are the fine embers, or firebrands, that have been documented in the Pacific Northwest and Australia, and that travel high into the atmosphere, drifting as far as 18.6 miles (30 km) in front of the fire. Those embers also create new spot fires.

There is nothing more terrifying than a monster fire with towering walls of flames that can travel as much as 14 miles (22.5 km) a day. These monster fires have been rightfully called firestorms, as they organize local wind patterns, stoking and fanning the fire.

The ponderosa pine ecosystem of the west is well adapted to fast moving surface fires. The natural fire frequency of lightning-induced burns is between 5 and 25 years.

Forests and Fire Suppression

For the past 18,000 years or so, people and fire have co-existed in Canada and the U.S. Paleo-humans hunted mega-ice-age mammals and they watched trees recolonize the land during the melting of the Pleistocene glaciers. Native North Americans set fires to promote greening-up of the grasslands to attract deer and elk for hunting. European settlers grazed livestock and harvested wood for fuel and construction. They also began to deliberately interrupt the fire cycle, which varies with terrain and elevation, by suppressing it.

In the 1930s, the U.S. Forest Service created a fire-suppression campaign around a bear called "Smokey." About a decade later, a scorched black bear cub from Lincoln National Forest in New Mexico became the real life Smokey Bear icon. Smokey's message, "Only you can prevent forest fires!" is still very valid today. Everyone must be careful in the forest — starting accidental fires is unacceptable and unlawful.

Prior to fire-suppression policies, the ponderosa pine ecosystem spread across 41 million acres (16.6 million ha) of western North America. Ponderosa thrive in drier ecosystems where they are often accompanied by Rocky Mountain Douglas-fir. Interspersed in these mid- to valley-bottom ecosystems are grasses. Lightning-induced fire frequently starts here, and these low-intensity surface fires race along the grasses. In parts of the vast ponderosa pine ecosystem, fire occurred naturally at intervals in the American southwest between two and seven years. Farther north, the intervals extended to about 25 years.

Both ponderosa and Douglas-fir have adapted to these surface fires by growing thick, insulative bark in excess of one foot (30 cm) and shedding the branches near the ground. Mature trees in these ecosystems hold their lowest branches 20 feet (6.1 m) above the ground so that the fast-moving surface fires cannot reach their foliage in the treetops. Fire that reaches the treetops is called crown fire. Crown fires are lethal in the ponderosa pine ecosystem. Surface fires clean the forest floor, exposing mineral soil where occasionally a tree seed may germinate, become a seedling, then a sapling, somehow avoiding the path of surface fires and eventually becoming a mature tree able to cope with fire licking at its lower trunk. Ponderosa pine ecosystems traditionally supported about 60 or 70 mature trees per acre (0.4 ha), as both ponderosa and

Douglas-fir can live well past 300 years and some to as much as 900 years.

Fire suppression over the past 70 years or so has significantly changed the composition and structure of the ponderosa pine ecosystem. Instead of 60 or 70 mature trees with dozens of saplings growing in a sea of grasses, there are now hundreds and in many cases thousands of saplings and young trees in an acre (0.4 ha) of forestland. When fire jumps back into these overcrowded forests, it ignites the foliage near the ground and kills the younger trees. More importantly, there are often many young trees and saplings growing underneath mature ponderosa and Douglas-firs and when they catch alight they act as a ladder, introducing fire into the crowns of the mature trees. Neither ponderosa nor Douglas-fir has evolved to contend with crown fires. The intense fires over the past 25 years or so in the ponderosa pine forests are killing the mature trees and removing a seed source for regeneration of burnt-over lands, and the surface fires, which are unnaturally hot due to very high fuel loads, are scorching the soils. The ponderosa ecosystem is not adapted to these stand-replacing monster fires.

Unlike the Douglas-fir and ponderosa pine, the lodgepole pine needs crown fires to thrive. The lodgepole pine forests cover 65 million acres (26.3 million ha) encompassing Alberta, British Columbia, Washington, Oregon, California, Montana, Idaho, Wyoming, and Colorado. All but the Californian populations depend upon massive stand-replacing crown fires. Lodgepole pines are specially adapted to fire because their closed cones hold viable seeds for up to 25 years. Once a fire's heat reaches about 145°F (63°C), the pine resin that holds the cone scales shut then melts and opens, releasing the seeds. The forest floor, having been quickly cleansed by a surface fire, is exposed, and the mineral soil can now receive millions of seeds from the burnt parent lodgepole pine trees. The disturbed site is quickly recolonized. A lodgepole pine forest, in which fire has a natural frequency of about 40 to 65 years, can live without fire for about 145 years (except in California, where trees can live for over 600 years). Lodgepoles begin to die naturally when the fire cycle is disrupted, or they become victim to voracious mountain pine beetles, as is happening currently throughout the lodgepole's range in the West.

After more than half a century of meticulous research, tree scientists and ecologists know that when they work with Mother Nature, and not against her, predictable outcomes will occur in the forests. That is, some fire must be allowed to occur on our landscape because fire is an important agent of change. And forests are always undergoing change. Smokey, too, has learned from scientists that we must have some fire in the forest. Otherwise, when fires eventually burn, they are burning with more intensity, ferocity, and severity than the ecosystems have evolved to cope with. As the West gets drier, we're going to see more fires.

The Okanagan Mountain Provincial Park Fire of 2003

The strong El Niño of the late 1990s resulted in tinder-like dry conditions across western North America. The year of 2002 was a hideous wildfire year across much of the West, and the following winter, snowfalls across coastal (excluding the Sierra Nevadas) and interior mountains ranges were far below normal. Spring meltwaters were meager. British Columbia, in particular, was caught in a long-lasting drought and as it entered the month of August 2003 — traditionally the worst fire month — hundreds of fires had already been raging across the province.

Millions of acres of dead or mountain pine beetle-infested lodgepole pine forests were kindling awaiting ignition. The largest native bark beetle infestation in modern times in conjunction with a prolonged drought caused all the forests throughout the province to be closed to timber harvesting or recreation — two multi-billion dollar industries.

On August 16, 2003, a wildfire was ignited by a bolt of lightning near Rattlesnake Island in Okanagan Mountain Provincial Park, adjacent to Kelowna, a city of about 96,000 people. The winds fanned the fire and it spread quickly, soon maturing into a firestorm.

The forests were dry throughout British Columbia. With 865 fires burning in the province, the British Columbia Forest Service needed additional help with the Okanogan Mountain Provincial Park fire. Fourteen hundred Canadian Armed Forces troops were called to action, and 60 fire departments stood shoulder to shoulder to fight the blaze. In six days, the fire grew to over 40,000 acres (16,188 ha) as it began a northerly and easterly march toward the city of Kelowna.

After one week, the raging firestorm began to consume dozens of homes built at the urban/wildlands interface surrounding Kelowna.

Thousands of people were evacuated from Kelowna and the surrounding areas. The fire destroyed a dozen homes as almost 1,000 firefighters worked around the clock to contain it.

Firefighters try and contain the Okanagan Mountain Park fire as it rips into the edges of the city of Kelowna, BC, in August 2003.

Maureen Curtis, a counselor for the Red Cross, tried to assist many of the homeless families and recalled, "The hardest thing for many is trying to figure out how to face their children, let alone the future."

At its zenith, the firestorm caused the evacuation of 45,000 people and consumed 239 homes. The final size of the Okanagan Mountain Provincial Park fire was 61,776 acres (25,000 ha). Miraculously, no lives were lost.

The fire kicked off an important debate: What is the role of fire in wildlands forests?

I had an active role in the media both in print and on television explaining that all forests have evolved around fire and that a blanket fire suppression policy in the West was not ecologically nor economically sensible or justifiable. Some fire must be allowed to occur on the landscape.

The California Firestorm of 2003

Each year, southern California experiences hot, dry winds, called Santa Ana winds, named after the Santa Ana Mountains and canyons from which they blow. These winds are sometimes deadly because they fan infernos. They can occur year round, but mostly take place between October and February. They are spurred on, mostly in December, by a huge high-pressure system over the Great Basin region that covers most of Nevada as well as parts of Idaho and Utah.

Winds as strong as 115 miles per hour (185 km/hr) pour out near canyon mouths. Typically, the strongest winds occur during the night and early morning hours when a sea breeze is absent. During the day, a sea breeze, which is an onshore wind, counteracts the Santa Ana winds.

The mechanics behind the Santa Ana winds are fascinating. Two pressure gradients, a high with clockwise winds centered over Nevada, Idaho, and Utah and a low with counterclockwise winds off southern California's coast, cause strong winds to shoot down the Santa Ana, San Gabriel, Santa Monica, and San Bernardino Mountains north and east of Los Angeles. Occasionally, upper-atmosphere turbulence fuels the upper-air winds, which results in even stronger Santa Ana winds.

Like much of the West in the summer of 2003, southern California was experiencing a prolonged drought exacerbated by voracious western pine beetles. The beetles had killed at least 10 million ponderosa, sugar (*Pinus lambertiana*), and Jeffrey (*P. jeffreyi*) pines, while another 5 million water-stressed, big-cone Douglas-fir (*Pseudotsuga macrocarpa*) and coulter (*Pinus coulteri*) and limber pines (*P. flexilis*) were dying or dead.

To make matters considerably worse, fire suppression over the past 70 years or so in the southern California mountain forests, in addition to a significant portion of the State, had promoted seeding-in of millions of acres of white fir (*Abies concolor*) and incense cedar (*Calocedrus decurrens*) seedlings and saplings. The traditional fire cycle every 20 years or so had kept the forest floor cleansed and the highly flammable foliage of white fir and incense cedar seedlings to a minimum.

Conditions were ideal for a fire: millions of drought-starved, dead or dying trees, millions more bark-beetle–killed trees, and tens of millions of highly flammable white fir and incense cedar seedlings and/or saplings unnaturally carpeting the forest floor. Add to this Molotov cocktail extremely dry, raging Santa Ana winds and a deliberately set rescue fire out of control, and in 2003 California experienced its largest firestorm in known times.

On October 25, 2003, a hunter got lost in Cleveland National Forest. Relying on his cell phone in a rugged terrain was futile. So at dusk he lit a signal fire. It escaped, and 40-mile-per-hour (64 km/hr) Santa Ana winds dispersed the ignition to tens of thousands of awaiting pine trees. The 20-acre (8-hectare) fire soon grew to 100 acres (40 ha).

The next day, October 26, another wildfire was reported 25 miles (40 km) northwest of Cleveland Forest. It grew very

The worst wildfires in California's history took place in fall 2003. A combination of Santa Ana winds, western pine bark beetles, and human-lit fires were a recipe for disaster. Fires burned into the tops or crowns of the trees and cyclonic winds stoked the flames into enormous firestorms overnight.

quickly, and began relentlessly spitting fireballs — firefighters' worst nightmare — half a mile (805 m) in front of itself.

A couple of days later, another fire took off from howling Santa Ana winds, exploding from 500 to 31,000 acres (202 to 12,546 ha) in four hours. Four wildfires converged in San Diego County, encompassing 12 percent of the entire county.

By day seven, southern California was ablaze: 16 fatalities, 2,400 homes in ashes, and 400,000 acres (161,874 ha) charred.

Steve Morphew of Escondido cried, "I have a home, but who cares — the fire took my wife."

Over 13 days, the Santa Ana winds pushed 13 scattered wildfires from the suburbs northwest of Los Angeles some 65 miles (105 km) into Ensenada, Mexico. Twenty-two people died and 4,000 homes were destroyed in southern California. The intense fires raged over 750,000 acres (303,525 ha) and caused in excess of $2 billion in damage.

CHAPTER 9
GLOBAL WARMING

The year 2005 was hot, the hottest ever recorded, with an average global surface temperature of 58.6°F (14.7°C). The winter of 2005–2006 was Canada's warmest on record. Between December and February, usually the coldest months, the country was 3.9°F (2.2°C) above normal. In 2006, a heat wave in July killed 200 people in the United States — 160 were from California. The first eight months of 2006 were the warmest January-to-August recorded in the U.S. In fact, the 11 hottest years on record have all occurred in the last 12 years. Overall, Earth's temperature is within 1.8°F (1°C) of its highest temperature levels in the past million years. What's going on?

In order to understand the magnitude of global warming, it is important to see how Earth physically works, including examining the cycle we are presently in, which is a warm period within an ice age cycle.

The Sun's Radiation

The sun provides Earth with its energy. Wind patterns and ocean currents disperse that energy globally. These two incredibly complex systems are intertwined and provide the basis for weather patterns.

The equator receives massive amounts of the sun's radiation. Air rises nine miles (15 km), forming an enormous clockwise cell called Hadley's cell (after Englishman George Hadley). Hadley's cell carries immense amounts of heat and moisture thousands of miles before sinking back down around latitude 30° north. Then between latitudes 30° and 60° north, another massive cell spinning counterclockwise rises nine miles (15 km), carrying heat and moisture thousands of miles farther. It's called the Ferrel cell (after American William Ferrel). Circulating clockwise four miles (6 km) overhead between latitudes 60° and 90° north is the gigantic Polar cell. It brings the warm air north, sinking it at the North Pole and bringing it southward at the surface. The southern hemisphere mirrors this three-celled system. Each respective southern cell circulates in the opposite direction from its northern-hemisphere counterpart.

Ocean currents, like their giant-geared wind counterparts, are also responsible for moving vast amounts of equatorial heat toward the poles. And they carry icy polar waters back toward the warm equator. The Gulf Stream in the Atlantic Ocean is an excellent example of this.

The warm, southerly Gulf Stream has been likened to a hot river moving through the Atlantic. It cruises along at 5.8 miles per hour (9.3 km/hr), and in sections it is 60 miles (96 km) wide and over 3,300 feet (1,005 m) deep. The heat from

another, more northerly, Atlantic current — the North Atlantic Drift — enables plants, animals, and people to live in the far northern parts of Europe, such as Lapland.

The driving factor that allows the warm Gulf Stream to circulate so far north in the Atlantic and then return southward is the difference in salt content between warm and cold ocean currents. Salty polar waters are heavier than warm equatorial waters. Near Greenland, a tremendous amount of sea water sinks toward the ocean floor. In essence, the frigid salty water pulls the warm Gulf water south again as though it were a weight on a giant conveyor belt. When this conveyor belt halts, conditions become right for the formation of ice ages.

In fact, all ocean currents are driven by this temperature/salinity difference called thermohaline circulation, and this accounts for sinuous, equatorial warm water mixing with cold polar water and its return.

The polar ice caps on Greenland and Antarctica are crucial, especially seasonally, because they supply icy meltwaters, which feed into oceans. The polar ice caps are also important because ice and snow help reflect over 85 percent of the sun's radiation. Together with the ice caps, the vast equatorial tropical clouds and desert sands reflect about 33 percent of the sun's incoming radiation to Earth back into outer space. This helps cool off our planet and maintain its temperature.

Global warming is significantly altering this equation. Vast amounts of both continental and sea ice are melting at an extraordinary rate. In 1979, satellite images showed that seasonal ice coverage equaled the size of continental United States — 1.7 billion acres (687 million ha). In 2002, the Larsen

B ice shelf of 1,268 square miles (3,284 km^2), about the size of Luxembourg, broke off Antarctica. In 2003, sea ice had shrunk by an area equal to Texas, Georgia, and New York, or 250 million acres (101 million ha). Between April 2004 and 2006, Greenland lost ice at about two and a half times the rate of the previous two years. That is the equivalent of more than 31 billion tons of meltwater per year.

About 288,000 square miles (745,000 km^2) of perennial Arctic sea ice, which normally does not melt during the summer, was lost from 2004 to 2005. Perennial sea ice can be 10 or more feet (3 m) thick. Over the past 150 years, the oceans have risen by between four and eight inches (10 and 20 cm), but in the last decade of the 20th century, the rate has alarmingly doubled.

The Greenhouse Effects

Radiation emitted by the sun passes through Earth's atmosphere, which resembles, in a crude way, a pane of glass on the roof of a greenhouse. As the sun's short-wave electromagnetic energy (visible wavelengths between 380 and 775 nanometers) passes through the glass roof, it hits the floor of the greenhouse — the surface of Earth. Energy is absorbed and then reradiated as a much longer wave (longer than 775 nanometers). The long-wave radiation cannot pass through the glass roof, and thus, the greenhouse gets warmer. Similarly, the next time you open your car door after it has been sitting out in full sunshine with windows closed, you will immediately notice that it is much hotter than the air outside.

The air in the lower section of the atmosphere, known as the troposphere, is made up of nitrogen (78 percent), oxygen (21 percent), argon (0.9 percent), and 0.1 percent other gases. This is the air we breathe. The "other" category includes gases such as CO_2, methane, water vapor, and about 27 other trace gases, which have been dubbed "greenhouse gases." Individually and acting in concert, the gases absorb and trap heat; as a result, they are raising the mean surface temperature on our planet, a phenomenon known as the "greenhouse effect."

Greenhouse Gases

Carbon dioxide is an essential compound enabling green life. It forms the base of the food chain, supporting all other life that exists on our planet. It makes up about 0.038 percent of the atmosphere or 381 parts per million, and it's rising. It takes about 1,000 years for the gas to be reabsorbed either by plants or the oceans. Carbon dioxide is the most prominent and high profile of the greenhouse gases. Carbon dioxide is very effective at trapping and holding atmospheric heat. Since the last Ice Age, the Pleistocene, some 14,670 years ago, levels of CO_2 have risen from 180 parts per million, and until 300 years ago they never exceeded 280 parts per million.

In 1958, on the top of Mauna Loa, on the big island of Hawaii, climatologist Charles Keeling began to monitor the amount of CO_2 in the atmosphere. He took samples of air to his laboratory and analyzed them. The graph from his findings shows a steady rise in CO_2 over the past 50 years from 318 parts per million to over 381 parts per million in 2006.

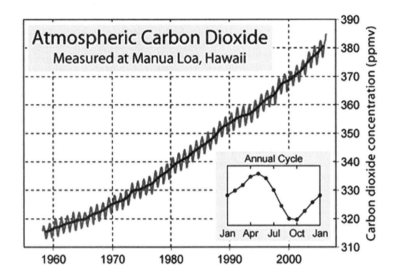

The saw-tooth effect on the chart is the result of the northern hemisphere trees awakening in the springtime and extracting CO_2 from the troposphere. Inside every leaf, CO_2 is used with water to make sugars in one of the most important chemical reactions that occurs on the planet. Trees and all plants and their skeletons are mostly made up of carbon. They are renowned for their ability to store it. Why then is this curve, called the Keeling curve, rising?

The remains of ancient green plants or "buried sunshine" lie beneath layers of rock or land or under the ocean. Over millions of years, the plant remains (made up mostly of carbon) have metamorphosed into coal, petroleum, and oil-based products. This is the main energy source that sustains humans in the early 21st century. When we burn these forms of energy, we release carbon back into the troposphere in the form of CO_2 gas.

Burning fossil fuels (coal, oil, and gas) released 23 gigatons of CO_2 in 2002. There are over 6 billion people on the planet. We are using, on average, four times more energy than our forebears did 100 years ago. Our consumptive use of fossil fuel has increased 16 fold over the same time period. We are now burning the equivalent of 422 years worth of sunshine locked in 300 million years of forests in just one year!

In addition, cutting down trees, or global deforestation, is adding significantly to rising CO_2 levels. Many of the deforested lands are deliberately set alight. Almost 30 percent of the CO_2 released into the atmosphere per annum comes from the burning of brushland for "slash-and-burn" subsistence agriculture.

Methane is the second most prominent trace gas in the troposphere at 1.76 parts per million. It has more than doubled in the last 200 hundred years. Methane comes from bacteria in swamps, rice paddies (rice is the largest staple crop in the world), and from cows' and other four-chambered ruminants' stomachs as they digest by belching and passing gas. Methane is also released when natural gas is burned. It is 23 times more efficient than CO_2 at absorbing heat, but thankfully it lasts only about a decade or so in the atmosphere before it breaks down.

Global warming is accelerating the melt of permafrost — soils that stay below freezing for at least two years — in northern Canada and Russia. In 2006, global warming caused methane to be released from the permafrost at a rate five times faster than previously predicted. Dr. Ted Schuur, an ecologist from the University of Florida, warns that rising levels of methane from the permafrost is "a slow-motion time bomb."

Water vapor is another trace gas in our atmosphere. As temperatures warm up, the troposphere is able to hold more water vapor. Water is extremely effective at trapping and holding heat. When CO_2 mixes with water vapor, particularly over the poles at high altitudes, their combined forces become formidable. The CO_2 slightly heats the troposphere, allowing it to hold more moisture, which then further warms the troposphere. In addition, high levels of methane, which is extremely efficient at trapping and holding heat, are preventing the troposphere from cooling at night. Essentially, the greenhouse gases are acting like the glass roof of the greenhouse, preventing our atmosphere from breathing freely at night.

Current levels of CO_2 are predicted to double by 2050. And even the conservative estimates predict that temperatures will rise by 5.5°F (3°C). The last time it was that warm was in the middle of the Pliocene, about 3 million years ago, when sea level was estimated to have been about 80 feet (25 m) higher than today. The result of global warming will be much more wild weather and significantly less available fresh drinking water.

The oceans are important not just because they disperse equatorial heat. They are incredibly productive ecosystems, absorbing a whopping two gigatons (billion tons) of CO_2 annually. And for every one molecule of CO_2 in the atmosphere, there are 50 in the oceans.

Forests and other green terrestrial life absorb another 1.7 gigatons of CO_2 annually. Saplings or young trees are important because they are experiencing a faster phase of growth than mature trees, so they need CO_2. The efficiency of land plants at withdrawing CO_2 will be somewhat offset as temperatures rise because soils in the far north are warming up. They are releasing methane and CO_2 into the atmosphere as decomposition rates accelerate.

The burning of fossil fuels is continuing to add at least 19.3 gigatons (from 2002 emission data) of CO_2 into the atmosphere despite the natural offset of both the oceans and forests.

Ice Ages

Ice ages are caused when the sun's radiation is reduced or becomes more variable. Remarkably, when solar radiation varies by just one-tenth of a percent, our planet's average annual temperature can rise or fall by an astounding 9°F (5°C). Ice ages result when solar radiation is reduced and temperatures fall, leading to an accumulation of miles-thick snow and ice in the continental ice sheets, polar ice sheets, and mountain glaciers.

There have been at least four major ice ages in Earth's past, beginning with the first approximately 2.7 to 2.3 billion years ago. Many cool periods have occurred during the last few million years, initially at a 40,000-year-average frequency but more recently at 100,000-year-average frequencies.

The Holocene

Atomic particles from oxygen in ice-core records reveal that Earth warmed abruptly 14,670 years ago. At the time, the amount of water locked in ice sheets and glaciers was staggering. The oceans were about 300 feet (91 m) lower. The melting that occurred from the eastern portion of the North

American ice sheet initially flowed south into the Mississippi River. Colossal ice dams jammed the flow south, and trillions of gallons of melt water were diverted into the St. Lawrence River, draining into the Atlantic Ocean.

Many scientists now believe that the immense amount of meltwaters shut down the Gulf Stream. Within less than a decade, temperatures on parts of Earth plummeted by 27°F (15°C). A small Arctic plant, *Dryas octopetala,* suddenly appeared in Scandinavia. Ice Age conditions set in 12,800 years ago and lasted 1,200 years before temperatures in Greenland abruptly rose 20°F (11°C) within a decade. This is considered the end of the last ice age — the Pleistocene — and the start of the Holocene, the name chosen for the current geological epoch that is classed as an interglacial, or warm, period.

From about 1500 to 1800 A.D., Europe dropped into a time called the "Little Ice Age." Cold temperatures occurred, especially between 1645 and 1715 A.D., a period called "the Maunder Minimum" after English astronomer Edward Maunder; the Thames and Dutch Ijselmeer Rivers regularly froze. A scarcity of sunspots — cooler patches or spots on the sun — during that time accounted for the much colder conditions. Oddly, decreased sunspot activity results in cooler temperatures on Earth. For example, during one 30-year period in the Maunder Minimum, only 50 sunspots occurred rather than the more typical 40,000 to 50,000 spots.

From his workshop in Cremona, Italy, Antonio Stradivari (1644–1737) made exquisite violins, violas, cellos, and guitars. The 1,100 or so instruments that he assembled using wood grown during the Maunder Minimum, including the most famous violin, the "Messiah" of 1716, produced unparalleled acoustics. The European spruce (*Picea* spp.), maple (*Acer* spp.), poplar (*Populus* spp.), and pear (*Pyrus* spp.) trees he used had unique wood properties. Tree growth during this cold period was very slow; as a result, the tree-ring growth was very narrow, and the wood cells were few but dense. Growing conditions like these have not occurred since, nor have any new violins or cellos been created with that melodious Little Ice Age snapshot.

The causes of the ice ages are complex. A number of theories have been proposed.

Milankovitch Cycles

A Serbian engineer, Milutin Milankovitch, calculated three distinct cycles on Earth that reduce the sun's radiation and cause ice ages. Remarkably, Milankovitch identified these cycles and published them in 1941, well before the modern age of computers.

Earth's orbital path is almost circular, but every 100,000 years it extends to slightly elliptical. The orbit can vary as much as 9.6 million miles (15.4 million km). When the orbit is at its farthest point from the sun, the intensity of solar radiation is greatly reduced, leading to a sharp drop in global temperatures.

The second cycle takes 42,000 years to complete and affects Earth's tilt or axis. The tilt of the axis can vary between 21.8 and 24.4 degrees. (Currently we are in the middle of this range.) The tilt of the equator changes during this cycle.

The third cycle occurs every 25,800 years. It has to do with Earth's precession, or wobble. It is caused by the fact that planet Earth is not a perfect sphere, but bulges outward at the equator. This cycle affects the intensity of the seasons.

When these three cycles combine, they minimize or maximize the solar radiation over hundreds of thousands of years. The dates Milankovitch calculated coincide with the ice ages.

Since we are currently in an interglacial or warm period with unprecedented rising CO_2 levels, it is imperative for scientists to continue to determine what the level of CO_2 and other greenhouse gases were during past interglacial times over the past 2 million years.

Buried Evidence

Paleo-scientists study the past. They are able to go back hundreds of millions of years and examine fossils. For example, by examining the tiny pores or stomates of fossilized plants, scientists can determine crude levels of atmospheric CO_2. The fewer pores, the higher the CO_2 levels. When a pore opens, it loses precious water vapor. Water is the lifeblood of Earth, and plants are very thrifty with its use. Higher levels of CO_2 translate into lower numbers of leaf stomates and correlate with warmer surface temperatures.

Deep-sea sediments that look like primordial ooze also give paleo-workers a glimpse at the past. These cores contain a wealth of information. For example, shell life contains special oxygen with neutral particles called neutrons in their atoms, and they decay at different rates depending upon whether they grew in warmer or colder seas. They are very accurate indicators of climate change. Likewise, different forms of carbon clearly show oceanic circulation patterns.

In addition to geological evidence, ice cores extracted

from beneath the Greenland and Antarctica ice caps reveal samples of the atmosphere stretching back 1 million years. From the evidence they have accumulated from different sources, scientists now have accurate historical records of atmospheric oxygen, CO_2, methane, and other trace gases. They are also able to determine when Earth experienced ice ages and the exact duration of the interglacial periods.

We have an in-depth record of the Holocene with cores from alpine lakes clearly showing layers of ash from lightning-induced fires over the past 12,000 years. Bristlecone pine trees from the White Mountains of east central California have enabled tree scientists and climatologists to precisely examine year by year the past 8,700 years. The tree rings clearly show years when moisture was more available (wider rings) as opposed to droughts (microscopic ring growth).

Of great interest to climate scientists and relevance for all humanity in the 21st century is the fact that when all the interglacial cycles are compared, there's only one warm period in the last million years when Milankovitch's cycles and temperatures were similar to those of today — in this case, about 1.8°F (1°C) warmer than they are today. It occurred about 430,000 years ago; that warm spell lasted for 26,000 years, as opposed to 12,000 years for all the other warm periods.

Does that mean we can expect another 14,000 years of warm weather before the onset of the next ice age? The answer to that question is unknown. The immediate concerns for a population that is belching 19.3 gigatons (and rising) of CO_2 into the atmosphere are the rate at which the surface temperature on Earth is rising and the accelerated melting of the polar glaciers. More specifically, how long before the vast quantity of meltwaters from the Greenland ice sheet shut down the Gulf Stream current? Between 1998 and 2004, scientists from Britain's National Oceanography Centre in Southampton have found that the volume of the current is beginning to slow down by as much as 30 percent. When the current stops, wild weather will be global and epic.

Climate Change Today

When climate scientists talk about change, they refer to it as being analogous to a finger on a switch. The system tends to jerk along to a point at which change very suddenly occurs. Climatologist Dr. Julia Cole refers to these events as "magic gates." Since 1970, she has pinpointed two crucial temperature events.

In 1976, a sustained, global increase in ocean surface temperature of 1°F (0.56°C) and a decline in the oceans' salinity by 0.8 percent was the first magic gate. Detailed records from 1945 to 1955 show that tropical Pacific Ocean surface waters often dropped below 66.5°F (19°C), but since 1976 they have rarely dipped below 77°F (25°C). The ensuing drought that followed in the Pacific Northwest as a result of elevated Pacific Ocean temperatures in 1977 was monumental, costing $500 million, causing algal blooms, and decimating the apple harvest by about 29 million boxes in Washington state.

The second magic gate occurred in 1998 with a fierce El Niño. El Niños result when the Pacific Ocean warms by as much as 9°F (5°C), unleashing droughts and floods that affect at least two thirds of the world. El Niño means "boy child" in Spanish, and is named for Christ because it occurs in December off the coast of Peru. El Niño events prevent

> **THE EL NIÑO PHENOMENON**
>
> El Niño is a major warming of the waters beginning in the western Pacific Ocean and drifting eastward. During an El Niño event, the affected area's winds weaken and sea temperatures become unusually warm. It typically results in a warm inshore current flowing along the coast of Ecuador, and about every seven to ten years it extends southward to the coast of Peru with frequently devastating effects on weather, crops, and fishing. Scientists now believe that El Niño has a return period of four to five years. It often lasts from 12 to 18 months and has repercussions around the globe.

nutrient-rich cold currents from upwelling, thereby denying food for anchovies, which migrate elsewhere, leaving the Peruvian fisherman to starve.

In 1998, the average global temperature rose about 0.5°F (0.3°C). The western Pacific Ocean temperature rose to over 85°F (29°C), which rerouted the jet stream over the North Pole. This caused a drought, which brought fire to 25 million acres (10.1 million ha) of southeastern Asian tropical rainforests. Naturally induced fires rarely occur in tropical rainforests, unlike in all other forest types. It is much more difficult for tropical rainforest trees to recolonize large, burnt-over, nutrient-poor soils compared to any other forest type.

Just as the polar ice caps hold the answers to the past, so today do they reflect the alarming speed and peril of global warming.

The staple of the food chain in the great Southern Ocean of the Antarctic is plankton. Plankton feeds krill; krill feed seals, penguins, albatrosses, and whales. Since 1976, krill have declined by 40 percent.

Krill are being replaced by small, jelly-like animals called salps, which until recently lived much farther north. Salps scavenge in ice-free waters. Whales do not eat salps, which have limited, if any, dietary value.

Over the last two decades, sea ice in the Southern Ocean has declined in area by 20 percent. Since krill depend upon sea ice, the decline has put the world's most productive ecosystem in peril, and this includes the largest creatures on the planet — whales.

Whales are not the only animals affected by the decline in sea ice. At the opposite pole, the great white polar bear rules the land. Unlike his grizzly cousins who can walk up and over mountains, polar bears travel over ice. They eat seals. And although polar bears have been documented catching seals in open Arctic waters, they mostly hunt on top of ice. Sea ice worldwide is dramatically melting, down 400,000 square miles (1.04 million km^2) over the past two-and-a-half decades. And not surprisingly, the numbers of polar bears is significantly diminishing. So, too, is the population of Arctic foxes, ravens, and ivory and Thayer's gulls, which all depend on the seal carcasses left by the polar bears. Summer Arctic ice is disappearing at about 9 percent per decade; at the predicted rates of warming, there will be no polar ice by 2040.

The Arctic Ocean is particularly sensitive to warming. The white ocean ice reflects most of the sun's heat back into space. However, as the melting occurs, white ice is replaced by darker ocean waters that absorb more heat energy, which in turn causes increased warming. In addition, the warmer it gets, the more permafrost that melts, in turn releasing more methane into the atmosphere. It's a process that scientists have aptly named a positive feedback loop.

The continuing rapid change in global ocean temperatures can be seen by the lack of sea ice in some areas. In July

2006, the southern tip of Ellesmere Island, just 932 miles (1,500 km) from the North Pole was ice-free. Usually the sea ice does not melt until the middle or end of August. The Arctic is once again hurtling toward another record-shattering hot year.

In fact, from 2000 to 2006 the Arctic has experienced six of the warmest consecutive years since record keeping commenced in 1880. Temperatures are at least 1.8°F (1°C) above average over the entire Arctic over the entire year.

Glaciers everywhere receded. Records over the past five years from stream gauges have shown that rivers flowing into the Arctic Ocean have been 3 percent to 9 percent higher than average with fresh melt water. The Arctic Ocean is becoming less salty, and in turn, this will disrupt the thermohaline circulation of ocean currents.

It is not just life at the poles that is showing the deleterious effects of global warming. Evidence can be seen everywhere on our planet. One thousand seven hundred plants and animals are migrating toward the poles at a rate of 4 miles (6.4 km) per decade, retreating up mountains 20 feet (6.1 m) a decade, and timing spring activities 2.3 days earlier per decade. That migration rate is not fast enough to keep up with the current rate of movement of a given temperature zone, which has reached about 25 miles (40 km) per decade in the period from 1975 to 2005.

Dr. James Hansen of the Goddard Institute for Space Studies and his team found that "rapid movement of climatic zones is going to be another stress on wildlife. It adds to the stress of habitat loss due to human developments. If we do not slow down the rate of global warming, many species are likely to become extinct. In effect, we are pushing them off the planet."

Clearly, many species cannot adapt quickly enough, so they are becoming extinct. At least 70 species of frogs, mostly mountain dwellers that had nowhere to go to escape the creeping heat, have gone extinct because of global warming. "Now we've got the evidence. It's here. It's real. This is not just a biologist's intuition. It's what's happening," says University of Texas biologist Dr. Camille Parmesan.

Fall 2006 was the warmest in central Britain since 1659 or the inception of record keeping. In Germany and Switzerland it was 10°F (5.5°C) warmer on average in November and the first half of December. From Norway to the Mediterranean, temperatures were also 5°F (2.7°C) warmer than normal. Austria's Central Institute for Meteorology and Geodynamics says it's the warmest it's been in 1,300 years. Snowfalls across Europe are half of what they were 40 years ago and winter is shorter. That winter was so warm that Geneva's chestnut (*Castanea* spp.) trees — the official harbinger of spring — were leafing. And fruit trees all over western Europe were in bloom. The fruit bearing flowers risk damage by winter frosts, which will result in poor fruit harvests in 2007.

Trees from Africa to Australia to the Rocky Mountains are feeling the effects of global warming. *Aloe dichotoma*, or the quiver tree of South Africa and Namibia — named by the San Peoples for its hollow branches that make excellent quivers — are dying off in parts of their range because they cannot adapt quickly enough to the climate changes. The spectacular king *Protea* — South Africa's national flower — and two-thirds of the Proteaceae family of 1800 species inhabiting Africa, Australia, and New Caledonia are forecasted to die off by 2050. The tenacious high elevation limber pines of the Rocky Mountain chain (from southwest Alberta to Mexico) are feeling the heat as a parasite, dwarf mistletoe, zaps their hard-earned sugars and adds additional stresses to many millions of water-starving pines.

Other species, like mosquitoes, are adapting and breeding later in the autumn. Scientists from the University of Oregon have shown that the mosquito *Wyeomyia smithii* from Alabama and Florida has begun to use global warming to drive its evolutionary clock. This is one critter that we should all fear.

The Effects of Global Warming on Wild Weather

As human-induced temperatures in the ocean and air continue to rise at an unprecedented level, predictions of more intense and severe weather systems are forecast. There is strong and mounting scientific evidence to support these predictions.

In 2005, the U.S. National Center for Atmospheric Research in Boulder, Colorado, documented a rising intensity of hurricanes in the North Atlantic. Massachusetts Institute of Technology found a 50 percent increase in the destructive power of cyclones (hurricanes, typhoons, and tropical cyclones) in the past half-century. Georgia Institute of Technology reported that over the past 30 years, the number of the strongest categories of hurricanes — Categories 4 and 5 — has doubled, and stronger cyclones are occurring in every ocean. The reason in all cases was definitive: rising sea-surface temperatures across the entire tropics. Global warming is heating the oceans.

Dr. Judy Curry of the Georgia Institute of Technology

says, "We can say with confidence that the trends in sea-surface temperatures and hurricane intensity are connected to climate change."

There are also strong links between human-induced global warming and rising ocean temperatures.

Dr. Robert Corell, chair of the Arctic Climate Impact Assessment, along with 18 other scientists, found that human activities accounted for two-thirds of the increase in water temperature in key hurricane-producing regions of the Atlantic and Pacific during the last century. "We've now learned that the human-induced buildup of greenhouse gases in the atmosphere appears to be the primary driver of increasing hurricane activity," says Dr Corell.

Experts from around the world warn us that we can expect more Katrina-like hurricanes as ocean temperatures continue to increase.

As Earth continues to experience stronger El Niños, characterized by unusually warm ocean temperatures in the equatorial Pacific, followed by La Niña's unusually cold ocean temperatures in the eastern equatorial Pacific, weather patterns will shift, and superceled mesoscyclones are predicted to intensify. Wild-weather tornadoes, like the strongest F6 ever recorded that hit Oklahoma City on May 3, 1999 with winds in excess of 318 miles per hour (512 km/hr), are predicted to occur again on the central plains, particularly during the La Niña cycles.

Global warming climate models have predicted that by 2050, the Rocky Mountain snowpack could be reduced by 60 percent in some regions, cutting summertime flow in half. The Colorado River, with its headwaters in Rocky Mountain

THE MOSQUITO

Of the conservative estimate of 10 million forms of all life on planet Earth, there currently exist 2,500 different kinds of mosquitoes. Despite being the size and weight of a grape seed, they are deadly and fearsome insects. Mosquitoes are benefiting from global warming because their habitat is expanding northwards from the equator. But the creature is also a master at adapting. How has something so tiny been able to successfully inhabit our planet for the past 80 million years?

It's all in the size and, in this case, it matters to be small. Mosquitoes are adapted to every terrestrial ecosystem from the tops of mountains to valley bottoms, from the Arctic Circle to the Sahara desert, and everything in between. They have thrived and adapted with the spread of human beings. In the last 300 years, the common house mosquito, which started in Africa, has gone global.

The mosquito is self-serving. She's not an aerator of soil like ants or worms, nor a pollinator like bees or moths, and not an essential food source for a particular species. Her goal is to feed and breed. Mosquitoes and the pathogens that they carry are extremely hardy. They are clever and relentless.

In 15 minutes, a female can lay about 250 eggs. One by one, they are symmetrically orientated headfirst and placed in a watercraft that has a pointy bow. All 250 float together in a war-canoe formation. Incubation takes about two days, and then the larvae hatch. Much longer, segmented, wriggly, whiskered, with specialized air tubes for breathing, they are set to grow and move into their next stage of development. Under magnification, they are as gruesome as any sci-fi monster. About 14 days after birth, they possess wings for moving, jaws for chewing, large compound eyes for finding their victims, a long spear-like drill for cutting through skin and sucking blood, and fuzzy antennae for smelling victims and sensing mates.

Their wings beat at about 250 to 500 strokes a second, they can attain and maintain speeds of 3 miles per hour (5 km/hr) and, rather clumsily, they can hover.

Their appetite for sex is insatiable, so two days after they become adult, they search for a mate. The process begins at dusk or dawn, when hundreds or perhaps thousands of males form a dancing swarm in the air near a landmark like a chimney or a church steeple. Males smell the females, lock together with them, and copulate. In many cases the fit is so tight that the male has some difficulties escaping and an unfortunate few manage to get away only by leaving their sex organs behind.

The female mosquito needs

THE MOSQUITO

just one ingredient to nourish her eggs — blood. She senses with her antennae the CO_2 and lactic acid that animals, including humans, exhale. This usually occurs at ground level because the scent plume is heavier than air and sinks to ankle level. Next, her compound eyes, similar to those of a house fly, locate the unsuspecting victim. She lands softly, and probes skin up to 20 times with her long, snake-like proboscis. Her salivary tube will deliver a chemical that inhibits the body's ability to stop any bleeding that might begin. Ninety seconds later, her body weight is three times what it was before feeding, and in an aeronautical feat, she just manages to sputter away.

In one of nature's most remarkable processes, within 45 minutes, she digests the blood by separating the water from the proteins and urinates pink droplets from her anus. The light solids are stored for creating future offspring. What the mosquito leaves behind in its saliva will either irritate the skin or potentially kill you. A female mosquito may live for about five months.

Yellow fever, malaria, dengue, encephalitis, and West Nile diseases have all, at one time or another, penetrated the southern United States. Each pathogen carried by loathsome species of mosquitoes is excruciatingly painful and, in some cases, lethal. Malaria kills millions of people each year. DDT, although toxic to our environment, is still the only known effective mass antidote to curtail the spread of malaria.

Instead of destroying this enemy, modern science is trying to genetically convert her. The race to find the genes to prevent pathogens from developing in mosquitoes is on. But in the meantime, human beings are dying by the millions as mosquito diseases expand worldwide at a frightening rate.

National Park provides water to the fastest-growing cities in the United States, including those in Arizona, California, Colorado, Nevada, New Mexico, Utah, and Wyoming. Over the past 50 years, the snowpack in the southern Rockies has melted two to three weeks earlier because of increasing springtime temperatures. "Water-use policies need to reflect not just what's going to happen in 2050, but what has already happened in the past 50 years," said University of Washington professor Philip Mote.

California is arguably the most remarkable geologic and botanical state in the union. The Sierra Nevada Mountains occupy one-fifth of the land in California and have a major influence on the climate, weather, and water supply of the state. Vast amounts of snow are intercepted and accumulated on the Sierra Nevadas during the winter. Currently, it is enough to supply 90 percent of the water needs for agriculture, industry, 34 millions denizens, and tens of millions of tourists each year.

Researchers from Stanford University and the University of California at Los Angeles forecast that global warming will cause much hotter summers and faster depletion of snowpacks, resulting in a reduced supply of water and food. They predict that by the end of the century, the snowpack on the Sierra Nevadas will decline at least 50 percent and that 75 percent of the high elevation forests will be lost.

Ecologists and tree scientists study patterns of vegetation across the landscape. After more than two decades of observations and studies, I have observed western North America becoming much drier.

For example, along the northwest edge of the Great

Plains grasslands, islands of white spruce trees are surrounded by a sea of grasslands. There are several reasons why the grasses are so successful in this ecosystem. They are experts at dealing with drought and fires because of their extensive and regenerative root systems. However, when trees appear in islands, that means there is slightly more moisture enabling them to dominate the grasses. Islands exist because of a seed source, and during the growing season at night mature trees are able to relax water held in tension in their trunks. They actively release some water at night through their roots to nourish smaller seedlings and saplings.

Over the past decade in southwestern Manitoba (in the center of North America), the drought-specialist grasses have been successfully invading and conquering the white spruce islands. The white spruce is in effect retreating northward. We know this because the number of new seedlings, called recruitments, is drastically lower. In other words the climate is becoming drier. During droughts, trees don't produce much seed, and the seed that is produced is unable to germinate because there is not enough moisture. In a last ditch effort after a mature tree has been water-starved, it will produce one last cone crop, called a distress crop, before dying. This phenomenon is visible along parts of the Great Plains grasslands.

The northern boreal forests of Alaska and the Yukon are also exhibiting extreme signs of global warming. These forests have adapted to cooler temperatures and plentiful moisture from a slow spring snowmelt and rainfall during the growing season. Tree-ring growth from white spruce in the interior of Alaska shows that growth occurs when temperatures are cool, and that in warm years, like those experienced in the past decade or so, the trees are shutting down. Global warming has stunted them.

From May to August, daily high temperatures have not changed much. However, there's been a huge upswing in daily lows; they increased by almost 5.5°F (3°C). "That's entirely consistent with the mechanism [by which] a greenhouse gas warming would operate. It operates to dampen heat loss. It doesn't add more heat so you get higher highs, but it dampens heat loss so you get higher lows," said University of Alaska forest ecologist Dr. Glenn Juday.

The average number of frost-free days in the interior of Alaska has increased from 80 days in the early 20th century to 120 frost-free days in the early 21st century. The greater number of 70°F (21°C) days, the loss of moisture available during the growing season, and warmer nighttime temperatures are causing the trees to stop growing during the summer — their normal growing time. As temperatures continue to rise in the far north, the forests will experience prolonged droughts, resulting in mass die-off. Drought-specialist grasses will successfully conquer the land.

As a matter of fact, that's exactly what's happening 30 miles (48 km) west of Whitehorse, Yukon, along the Alaska Highway. A forested valley that burned in 1958 has reverted to grasslands with only a scant cover of aspen trees.

Another much more massive and equally alarming phenomenon in the forests has been taking place across the entire West, from the mountains of New Mexico north to the Yukon, from Arizona to California, from Utah to Idaho, and from Oregon north to Alaska. There are four indigenous bark beetles — spruce beetles, mountain pine beetles, western pine beetles, and engraving pine beetles — eating their way through billions of mature forest trees.

Why are the forests of the West exploding with these native and lethal insects? What is going on?

In part, the answer lies in forest-management policies. We use wood and wood products in our society, and therefore the demand is high for them. So our current policies in North America, particularly in the West, favor and promote fire suppression. We have deliberately and successfully interrupted the natural fire cycle. Many of our natural forestlands are overcrowded and stressed.

On the other hand, our exquisite and diverse valley-bottom, mid-slope, and high-elevation forests of the West are susceptible like other forms of life to droughts and global warming.

What we are seeing across the entire broad spectrum of forests is a classic response to drought by opportunistic bark beetles as they are altering the landscape. Forest ecosystems are very dynamic communities that respond to change. Change can be as small as a single tree falling over, a patch of trees being infected by a root disease and being blown down, strong coastal winds flattening a hillside of trees, a lightning-induced fire burning a thousand acres, or an insect infestation.

In the case of the western forests, billions of trees have undergone prolonged water stress since the late 1990s. When pines and spruce are water stressed, they stop producing their natural defense mechanism: gooey pitch. In addition, water-stressed trees emit a chemical distress message that is detected by bark beetles.

Bark beetles land on tree bark and bore through it, stopping just inside the living bark, which is very rich in sugars that support tree life. Once inside, they eat, lay eggs, and move onto other helpless trees. In the meantime, the beetle has introduced a lethal fungus into the conducting part of the trunk, called the xylem, which carries water to its top or crown. The fungus proliferates in the conducting tracheal cells and prevents water from moving upwards. Essentially, it plugs the tree's plumbing, and the tree dies from water starvation. There is, however, enough stored sugar on the inside of the bark, called the phloem, for the developing larvae, either later in the summer or in the springtime, to eat, mature, and disperse.

There are many natural factors that can keep beetle populations in-check: fire and cold temperatures are the main two environmental factors.

The fire cycle has been, for the most part, successfully interrupted and removed from the equation. Global warming has had a fantastic effect on these prodigious bugs. Spanning from latitude 32° to 62° north, four species of tree-eating beetles have sped up in a very short space of time (about a decade) their reproductive cycle, particularly the mountain pine and spruce beetles. Historically, these two species required two years between breeding cycles. Spruce beetles are now breeding each year in Alaska and the Yukon, while the mountain pine beetles of southern British Columbia are breeding, in some cases, twice a year.

Global warming has significantly impacted the winters in the north. Bark beetles are susceptible to lethal November temperatures of –40°F (–40°C) for a period of at least 72 hours.

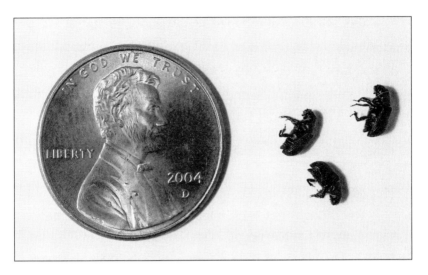

Dead bark beetles are displayed next to a penny at Boise State University in Boise, Idaho. Four species of native bark beetles have killed at least 1 billion trees across the West. They are all roughly the same size.

These temperatures normally occurred in central British Columbia and the Yukon, but during the past 15 years they have not occurred.

As a result, British Columbia is experiencing the largest bark-beetle infestation ever recorded in modern times. Bark beetles are spilling over from British Columbia into Washington, Idaho, Montana, Alberta, the Yukon, and the Northwest Territories. In the far north, Jack pines have not evolved to cope with these insects. Jack pines will provide a vast new food source for expanding beetle populations to spread possibly as far as Labrador. The mountain pine bark beetles have destroyed enough timber in British Columbia alone to supply the entire United States housing market for three-and-a-half years.

From British Columbia to New Mexico and almost every point in between, bark beetles are feasting, breeding, and flying. The state tree of New Mexico is the thrifty pinyon pine (*Pinus edulis*). Millions of pinyon pines and their constant companions, junipers (*Juniperus spp.*), dot the landscape of the southwest, forming clumps of open or woodlands forests, comprising the fourth-biggest ecosystem in America. They exist in a semi-arid environment with about 13 inches (33 cm) of precipitation a year.

The drought has been relentless in New Mexico. Ninety percent of the pinyon pines of New Mexico — millions of trees — are dead. The drought weakened the trees, and bark beetles moved in for the kill. Life in the southwest is precariously perched on the edge. Global warming has increased the annual temperature by only 2°F (1.1°C), just enough to push pinyon pine over the edge with a helping hand from the beetles. When the woodland ecosystem disappears, so do all the critters. Tree scientists and climate experts are rightfully concerned. Will our species be next?

One of the main concerns about global warming is droughts and how they will affect fresh drinking-water supplies. So far, we've seen that snowpacks have diminished in the Cascade Mountains and southern Rocky Mountains, and predictions over the next 80 years or so for the Sierra Nevadas are also grim. The continent of Australia is drying out at an even more alarming rate; drinking-water supplies for about 5.5 million people in Perth and Sydney are in jeopardy.

Global warming will increase the incidence of droughts and continue to reduce the available fresh drinking water. Will our society be able to cope with droughts, or will the same fate that befell the Akkadians, Mayans, and Anasazi peoples strike us? Another valid concern is that droughts promote wildfires. A recent study found that global warming is in fact fueling mega-wildfires, particularly in the northern Rocky Mountains. Warm springs are melting the snowpacks much earlier than normal, the forests are drying out, and large wildfires are happening when this occurs. "When you have a warm spring and early summer, you get earlier snowmelts. With the snowmelts coming out a month earlier, areas then get drier overall and there is a longer season in which a fire can be started. There's more opportunity for ignition," remarked Dr. Anthony Westerling of the Scripps Institute of Oceanography.

That's exactly what happened in Montana in August 2006 with the 208,000-acre (84,170 ha) Derby Mountain fire that burned for a month and destroyed 29 homes. The area that burned in Montana in 2006 was in excess of 940,000 acres (380,400 ha), and scientists have predicted that the number of acres that burn in that state could increase five-fold by the end of the century.

"Lots of people think climate change and the ecological responses are 50 to 100 years away. It's happening now in forest ecosystems through fire," says Professor of Dendrochronology Thomas Swetnam of the University of Arizona.

Global warming is also pushing fires in the West higher up into mountain ecosystems that burn far less frequently than valley bottoms. Over the past two decades, those areas, ranging from 5,300 to more than 8,000 feet (1,615 to 2,440 m), have had the largest increase in big fires.

As temperatures on Earth rise due to global warming,

more wild weather like droughts will spur on more wildfires, thus fueling firestorms across the West and elsewhere. In addition, because of fire suppression, bark-beetle populations across western North America have run amuck.

The sun is Earth's powerhouse, and the ocean currents and wind patterns are essential to dispersing equatorial heat. Equally important are the polar ice caps, sea ice, and land glaciers because they, too, help keep temperatures regulated on the planet. Our atmosphere has been likened to a greenhouse, and accelerated burning of fossil fuels and emissions of CO_2 have caused surface temperatures on Earth, since record keeping began, to rise significantly. Rising temperatures are imposing stressful and life-threatening conditions for terrestrial and aquatic ecosystems around the globe. Humans depend on the natural world for drinking water, food, and clean air. Rising temperatures also affect weather patterns, and in turn this affects the way we live.

CHAPTER 10
HOPE FOR THE FUTURE

There is a plethora of incontrovertible scientific evidence from around the world showing that Earth is warming up at an unprecedented rate. Plants and animals are dying because their habitat is disappearing. Climate scientists, too, have predicted more, and more intense, wild weather around the globe.

Global warming is also becoming an enormous political and economic problem. It is causing global security issues. In Darfur, Sudan, an intense drought has caused pastoralists and farmers to fight over any lands where rain falls.

"There will be more Darfurs," warns British foreign secretary Margaret Beckett.

Sir Nicholas Stern, a senior British government economist, predicted that unchecked global warming would devastate the world economy. It will cut the standard of living by as much as 20 percent and plunge the world into a recession worse than that of the 1930s.

The cost of natural disasters exceeded $225 billion in 2005, up from the previous record of $118 billion in 2004, according to reinsurance behemoth Swiss Re. A top executive was quoted as saying, "Global warming has accelerated from a problem that might affect our grandchildren, to one that could significantly disturb the social and economic conditions of our lifetime."

The price of gasoline in the United States is nearing $3 a gallon and rising, still less costly than elsewhere in the western hemisphere, but it has almost doubled at the pump in the past 18 months, and it's causing the insatiable U.S. economy some indigestion pains as it wrestles to keep inflation to a minimum.

In this one case, the indigestion pains may be good. It is quixotic to expect that the global economy currently driven by coal, oil, and gas will change overnight to an alternative fuel source, irrespective of the known dire consequences of fossil fuel consumption, namely elevated temperatures, melting ice caps, rising ocean levels, and more catastrophic wild-weather events.

There is, however, real hope on the horizon.

Three-dollar-a-gallon gasoline is nearing a tipping point that will benefit the environment. Although Toyota, Honda, and Ford's hybrid cars currently make up an infinitesimal percentage of the enormous North American car fleet, consumers are flocking by the thousands to purchase hybrid cars as quickly as they are produced.

The change we are about to see, interestingly, is "eco"

driven — not *eco* as in *ecology*, but as in *economy*; that is, due to economics of rising fuel prices. And this is just the beginning.

Change is all around us, and it is not limited to the automotive industry. As a matter of fact, two oil giants, British Petroleum and Royal Dutch Shell, have shown sound leadership and are spending millions of dollars per annum in researching alternative, renewable forms of energy, including substantial investments in wind and solar power rather than their current bread and butter — oil and gas.

On March 1, 2006, the Aspen Skiing Company, which operates four Colorado ski mountains and two hotels, announced that 100 percent of all their energy requirements to operate would be derived from wind farms. "Clearly the most pressing issue of our time is climate change, and addressing energy use is one of the most important actions we can take on that front," said president and CEO Pat O'Donnell.

Whole Foods Market has 184 organic supermarkets in North America and Great Britain, and 100 percent of all their energy for all their stores is sourced from alternative energies, including geothermal, small-hydro, solar, biomass, and wind turbines.

DuPont, a mammoth chemical company, has already reduced its greenhouse gas emission by 72 percent since 1990. They are developing and marketing an array of "green" products, including Tyvek insulation, energy-efficient refrigerants, and corn-based raw materials that will replace plastic produced from oil.

HSBC North America derives 35 percent of its energy to run their mighty banking empire from wind turbines.

Pharmaceutical and health-care giant Johnson and Johnson uses biomass, small-hydro, solar, and wind power to offset 30 percent of its energy needs. The titan of the coffee world, Starbucks, relies on wind turbines to provide 20 percent of its energy needs. Computer chip manufacturer Advanced Micro Devices runs its entire Austin, Texas, facility on biogas and wind power. And Intel, the biggest chip maker in the world, is the largest purchaser of wind power in both Portland, Oregon, and in the entire state of New Mexico.

The growing list of blue-chip companies in the United States and elsewhere that are striving to reduce greenhouse gases is being bolstered monthly as more companies begin to distance themselves from 19th-century coal, oil, and gas technologies.

General Electric is leading the pack with its cutting-edge technologies in a number of different sectors. They build more wind turbines than anyone else. They have also committed to double their research budget from $700 million to $1.5 billion annually for the next five years for cleaner, more efficient locomotives and jet engines, and to enhance existing technologies of solar power and wind turbines. "We think green means green. This is a time period where environmental improvement is going to lead toward profitability," says Jeffrey Immelt, Chairman and CEO of General Electric.

Wal-Mart has 6,600 stores worldwide; they lead not only in retailing, but also by example. CEO Lee Scott says, "One little change in product packaging could save 1,500 trees. If everyone saves 1,500 trees or 50 barrels of oil, at the end of the day you have made a huge difference." Wal-Mart is making tremendous inroads by beginning to implement solar panels on their store roofs, coating their exterior walls with heat-reflective paint, and using high-tech systems that automatically dim or raise lights depending on whether it's sunny or overcast. In Texas, Wal-Mart has begun trials of wind turbines in their parking lots. Not only is this benefiting the environment, but Wal-Mart expects to save $310 million a year in energy costs.

It's not just corporate America that is seizing the opportunity to rethink its energy use and make sound economic changes to cut CO_2 emissions. Two hundred and eighteen mayors in 39 states representing 44 million Americans have bought into making a difference, and they are embracing and implementing new, clean, alternative-energy technologies.

Mayor Will Wynn of Austin, Texas, said, "We're frustrated by the lack of national leadership. This is about the future of the planet." Seattle's Mayor Greg Nickels launched the U.S. Mayor's Climate Protection Agreement and his philosophy is, "If it's not going to happen from the top down, let's make it happen from the bottom up."

Governor Arnold Schwarzenegger of California signed into law the Global Warming Solutions Act of 2006, making that state the first to impose a universal cap on greenhouse-gas emissions. The act mandates the reduction of greenhouse gases by 25 percent by the year 2020. Governor Schwarzenegger has challenged the Bush administration and Congress to follow, saying, "California will lead the way on one of the most important issues facing our time."

Private foundations and individual philanthropists are also joining forces to assist in finding solutions to rising global temperatures. For example, Sir Richard Branson, entrepreneur, adventurer, and billionaire, pledged $3 billion to the Clinton Global Initiative over the next decade to combat global warming by developing new, alternative energy sources.

About one-third of Earth is covered in sun-rich deserts, a potentially vast source of energy. If four percent of those sun-rich deserts are harnessed with solar panels, the current world's energy needs would be met. Sanyo Electric has a model for a global energy infrastructure that would harness the sun and interconnect the whole world's electric grids through super-efficient, high-capacity, intercontinental transmission lines.

Dinosaurs ruled Earth for 200 million years. Humans have been on the planet for about 7 million years. The dinosaurs were brawny; our species is gifted with a large brain. If we can plan to put a person on Mars, then we should easily be able to reduce greenhouse gas emissions by clearly thinking our way out of accelerating global warming. Twenty-five years ago, recycling was in its infancy. Today, recycling is a multi-billion-dollar worldwide industry. Reducing greenhouse gas emissions by using alternative renewable energy will also follow the lucrative recycling paradigm. There is a lot of hope, and our future is riding on it.

WILD WEATHER TIMELINE

430,000 B.C. The interglacial or warm period lasted 26,000 years.

14,670 B.C. The oceans were 300 feet (91 m) lower than present day. Earth warmed abruptly, and the CO_2 in the atmosphere was 180 parts per million.

12,800 B.C. Earth's temperatures plummeted 27°F (15°C), a change that lasted for 12 centuries.

11,600 B.C.. Greenland's temperatures rose abruptly by 20°F (11°C).

6,700 B.C. The beginning of continuous climate records from bristlecone pines in White Mountains of California.

4,200 UNTIL 3,900 B.C. A massive drought in present day Syria and Iraq decimated the Akkadian Empire.

810, 860, AND 910 A.D. Three-, six- and nine-year droughts in present day Mexico along the Yucatan peninsula wiped out millions of Mayans.

1091 — OCTOBER 17 A tornado obliterated 600 homes and many churches in England.

1276 TO 1299 Massive drought annihilated Anasazi people of the Four Corners region in the American southwest.

1645 TO 1715 This period, called the Maunder Minimum, was dubbed the Little Ice Age due to lack of sunspots and freezing temperatures in Europe. CO_2 levels were 280 parts per million.

1888 — JANUARY 12 During the Children's Blizzard in the Dakotas and Nebraska, over 100 children perished in freezing conditions.

1900 — SEPTEMBER 8–9 Galveston, Texas, hurricane killed at least 8,000 people.

1947 — JANUARY 31 TO FEBRUARY 10 Canada's longest blizzard, it lasted 10 days and nights.

1953 — MAY 11 Waco, Texas, was hit by an F5 tornado with winds gusting at 285 miles per hour (459 km/hr), killing 114, injuring 1,097, and causing $51 million in damages.

1958 Climatologist Charles Keeling began to monitor CO_2 on Mauna Loa, Hawaii. CO_2 levels at 318 part per million.

1998 — JANUARY 5 TO 10 Worst ice storm in modern times blanketing eastern Ontario, Quebec, western New Brunswick, Maine, New Hampshire, Vermont and upstate New York. About 74,560 miles (120,000 km) of power lines and telephone cables, 130 major transmission towers, and 30,000 wooden utility poles were destroyed.

2002 This year marked the driest year in 1,000 years in the state of Arizona.

2003 — AUGUST 16 Okanagan Mountain Park fire in British Columbia consumed 61,776 acres (25,000 ha), destroying 239 homes.

WILD WEATHER TIMELINE

2003 — OCTOBER 25 Firestorm in southern California lasted 13 days, leaving 22 people dead, 4,000 homes destroyed, 750,000 acres (303,515 ha) charred, and $2 billion in damage.

2005 — AUGUST 28 Hurricane Katrina came ashore, pummeling 6,400 miles (10,300 km) of Gulf coast, demolishing hundreds of thousands of cars and 75,000 boats in Louisiana alone. It also resulted in five major oil spills, 1,836 dead, and $75 billion in damages.

2006 Australia experienced its worst drought in 100 years. Australian wheat, barley, and canola crops were decimated; losses were estimated at $6 billion.

2006 Great Britain shattered a number of records, including the warmest temperature ever recorded in July at Wisley — 97.7°F (36.5°C) — and the warmest fall ever recorded since records began in 1659. It was also the warmest autumn in the Netherlands since 1706 and since 1768 in Denmark. Canada experienced its mildest winter and spring on record and the United States had its warmest January–September ever. Forest fires across the country consumed almost 10 million acres (3.9 million ha).

2006 Severe drought affected more than 10 million people in the Greater Horn of Africa, but the region also suffered some of the worst floods in 50 years. Drought reduced 11 percent of the soybean crop in Brazil and damaged millions of acres of crops in China.

2006 CO_2 levels measured at 381 parts per million.

AMAZING FACTS AND FIGURES

THE WORLD'S DEADLIEST single tornado struck on April 26, 1989. It killed 1,300, injured 12,000, and left 80,000 homeless in towns of Saturia and Manikganj, Bangladesh.

ONE HUNDRED TONS of ancient plant life is required to create 1 gallon (3.8 liters) of gasoline.

NATIVE PINE AND SPRUCE bark beetles have destroyed enough timber to supply 5 years of wood to the U.S. housing market.

THE STRONGEST TORNADO WINDS ever measured reached 318 miles per hour (512 km/hr) in Oklahoma City on May 3, 1999.

REPORTS AROUND THE WORLD of it raining frogs, fish, mussels, and pond turtles have been verified. They result from waterspouts sucking the critters out of their aquatic ecosystems and spitting them out on land. The saying that it "rains cats and dogs," however, bears no merit, though references date back to Norse mythology and medieval England. Likely it referred to torrential rainfalls that drowned malnourished cats and dogs.

THE COLDEST RECORDED North American temperature is −81.4°F (−63°C) at Snag Aerodrome, Yukon, on February 3, 1947.

THE MOST SNOW in a single snowstorm was 15.75 feet (4.8 m) occurring at Mt. Shasta Ski Bowl in California between February 13 and 19, 1959.

YUMA, ARIZONA, is the sunniest place in the world, receiving 4,055 hours each year out of a possible 4,456 hours.

AT ANY ONE PARTICULAR TIME there are approximately 1,800 thunderstorms occurring in Earth's atmosphere.

THE WORST HURRICANE ever recorded was Katrina, costing $75 billion.

MT. WAIALEALE in West Maui has over 350 days with rain per year, receiving 460 inches (1,168 cm) of rainfall.

THE STRONGEST WIND GUST ever recorded on Earth was 231 miles per hour (371 km/hr) on Mt. Washington, New Hampshire, April 12, 1934.

WHAT OTHERS SAY

"In the 1970s, there was an average of about 10 Category 4 and 5 hurricanes per year globally. Since 1990, the number of Category 4 and 5 hurricanes has almost doubled, averaging 18 per year globally."

DR. PETER WEBSTER, PROFESSOR AT GEORGIA INSTITUTE OF TECHNOLOGY'S SCHOOL OF EARTH AND ATMOSPHERIC SCIENCES.

"If we carry on with business as usual in all likelihood three out of every five species will not be with us at the dawn of the next century."

DR. TIM FLANNERY, PROFESSOR AT THE UNIVERSITY OF ADELAIDE.

"The way democracy is supposed to work, the presumption is that the public is well informed. The U.S. government is trying to deny the science behind global warming."

DR. JIM HANSEN, DIRECTOR OF THE GODDARD INSTITUTE FOR SPACE STUDIES.

"The middle of the United States and certainly the Southwest are well on their way to one of the worst droughts in history."

DR. CARL ANDERSON, PROFESSOR AT TEXAS A&M UNIVERSITY.

"For us as a company, the debate about CO_2 is over. We've entered a debate about what we can do about it."

JEROEN VAN DER VEER, CHAIRMAN AND CEO ROYAL DUTCH SHELL.

"America's oil and coal industries receive more than $20 billion a year in subsidies. Imagine what could be done if that money was invested in green energy."

ROSS GELBSPAN, PULITZER-PRIZE-WINNING JOURNALIST

"The apparent sensitivity of ice sheets to a warmer world could prove disastrous. The greenhouse gases that people are spewing into the atmosphere this century might guarantee enough warming to destroy the West Antarctic and Greenland ice sheets. That would drive the sea up 16 to 33 feet [5 to 10 m] at rates not seen since the end of the last ice age. This will push a half a billion people inland. And this is not an experiment you get to run twice."

DR. MICHAEL OPPENHEIMER, PROFESSOR AT PRINCETON UNIVERSITY.

SELECT BIBLIOGRAPHY

Flannery, Tim. *The Weather Makers.* New York. Atlantic Monthly Press, 2005.

Halter, Reese and Turner, Nancy. *Native Trees of British Columbia.* Banff and Rancho Mirage. Global Forest Society. 2003.

Kricher, John. *The Ecology of Western Forests.* Norwalk, CT. The Easton Press. 1993.

Lanner, Ronald. *Conifers of California.* Los Olivos, CA. Cachuma Press, 2002

Larson, Erik. *Isaac's Storm.* New York. Random House, 1999.

Laskin, David. *The Children's Blizzard.* New York. Harper Collins, 2004.

Marchand, Peter. *Life in the Cold.* Hanover, NH. University Press of New England, 1991.

Weart, Spenser. *The Discovery of Global Warming.* Cambridge, MA. Harvard University Press, 2003.

Weems, John Edward. *The Tornado.* College Station, TX. Texas A&M University Press, 1977.

Web Sites

National Oceanic & Atmospheric Administration's paleoclimate site: *www.ngdc.noaa.gov/paleo*; Global Change Research Information Office (U.S. government research): *www.gcrio.*

org; News items from U.S. Global Change Data and Information System: *globalchange.gov*; National Interagency Fire Center: *www.nifc.gov/*; U.S. National Assessment on Climate Change: *www.usgcrp.gov/usgcrp/nacc;* Intergovernmental Panel on Climate Change: *www.ipcc.ch*; SafeClimate for Business: *www.safeclimate.net;* The Discovery of Global Warming (climate change history): *www.aip.org/history/climate*

PHOTO CREDITS

Cover: The Image Bank/Pete Turner; AP Photo/Troy Maben: page 141; Clipart.com: page 135; CP Images/Jacques Boissinot: page 83; Creatas: pages 26, 34, 64, 90, 94, 103, 108, 111; Digital Vision: pages 21, 30, 31, 45, 48, 63, 84; FEMA: pages 12 (Jocelyn Augustino), 13 (Andrea Booher), 16 (Jocelyn Augustino), 39 (Andrea Booher); NASA: page 93.

ACKNOWLEDGMENTS

The author acknowledges the following sources for quotes included in this book:

Caribbean Coral Dying at High Rates, Associated Press, March 31, 2006. *Isaac's Storm,* Random House, 1999. Storm Warning, *New Scientists,* September 24, 2005. *Perils of a Restless Planet,* Cambridge University Press, 1999. *The Tornado,* Texas A&M Press, 1977. Edmonton Tornado *www.cbc.ca/ar-*

chives, 1987. *The Children's Blizzard,* Harper Collins, 2004. Ice Storm Devastates Parts of Canada and US Northeast *www.cnn.com*, Jan 10, 1998. *Two Mile Time Machine,* Princeton University Press, 2001. Aspen Ski Company Goes 100 Percent Green, *www.usepa.gov,* March 1, 2006. The Tipping Point, *Time,* April 3, 2006. Fortune, March 21, 2006. Relief Centers Give B.C. Fire Evacuees Shelter, *www.ctv.ca,* August 23, 2003. Firestorms 2003:The Story of a Catastrophe, *www.nctimes.com,* November 15, 2003. A Warming Trend of Less Ice, *Science,* 24 March 2006. Dry Southwest in the Line of Fire, *LA Times* June 25, 2006. Drought in US, *LA Times,* June 28, 2006. The Weather Makers, *Atlantic Monthly Press.* Global Warming, Inter Press Service, September 21, 2004. *An Inconvenient Truth,* Rodale. *Path of Destruction,* Little, Brown & Co. Climate Change to Hit WA Wine *www.thewest.com.au,* Oct 6, 2006. Warming Reshapes Alaska's Forest, Associated Press, Sept 11, 2006. Warming Climate May Increase Western Wildfire Woes *www.scientificamerican.com,* July 7, 2006. California Takes the Lead in Global Warming Fight, Associated Press, Sept 28, 2006. Global warming already killing species *www.cnn.com,* Nov 21, 2006.

ABOUT THE AUTHOR

Dr. Reese Halter is an award-winning conservation scientist, family man, best-selling children's author, syndicated science writer, TV nature documentary host, and professor of Botany at Humboldt State University, California.

Dr. Reese's love of nature began as a child. A springtime

tree-planting ritual with his father and brother became his passion. He knew from the time he was a child that he wanted to be a tree scientist and went on to attain three university degrees, including a PhD from The University of Melbourne, Australia.

It became clear at a young age to Dr. Reese that there was a tremendous lack of basic information on how trees and forests function. He believed that teams of multidisciplinary problem-solving scientists needed to work together to short-circuit ecological disasters, and identify and protect fragile ecosystems.

In the late 1980s, Dr. Reese founded Global Forest Science as a charitable international forest research foundation. He donated the first seed money to the foundation. Today, with an international team of over 140 scientists, Global Forest Science is a world leader in forest science research and conservation and has been called the Red Adair of the forest biology world. Global Forest Science has many victories, including the legislation from Ottawa to protect the threatened westslope cutthroat trout of British Columbia and Alberta, protection of the world's largest ant colony in Japan, using trees and forests in Manitoba and Wyoming as a barometer of rising global temperatures, opening an international insect quarantine facility at Simon Fraser University in British Columbia, saving New Zealand's multi-billion-dollar forestry and agriculture industries from the Australian painted apple moth, and understanding dieback of the tallest trees on Earth — California redwoods.

Through Global Forest Science, Dr. Reese visits schools and encourages children worldwide to embrace conservation, science exploration, and learning. For more information visit the author's web site at *www.DrReese.com*.